ELECTRICITY IN THE HOME

G. Davidson
C.Eng., F.I.E.E.
and
L. C. Lamb
C.Eng., F.I.E.E., M.C.I.B.S.E.

D1465029

TEACH YOURSELF BOOKS

Hodder and Stoughton

First edition 1945
Second edition 1958
Third edition 1973
Fourth edition 1983
Fifth edition 1989

Copyright © 1958 edition
Hodder and Stoughton Ltd
Copyright © 1973 and 1983 editions
G. Davidson
Copyright © 1989 Revised edition
L. C. Lamb

All rights reserved. No part of this publication may be
reproduced or transmitted in any form or by any means,
electronic or mechanical, including photocopy, recording,
or any information storage and retrieval system, without
permission in writing from the publisher.

British Library Cataloguing in Publication Data
Davidson, G. (*George 1901–   *)
Electricity in the home. 5th ed.
1. Residences. Electric wiring systems – Amateurs'
manuals
I. Title   II. Lamb, L. C.
621.319′24

ISBN 0 340 49703 3

Printed in Great Britain for
Hodder and Stoughton Educational,
a division of Hodder and Stoughton Ltd,
Mill Road, Dunton Green, Sevenoaks, Kent,
by Richard Clay Ltd, Bungay, Suffolk
Photoset by Rowland Phototypesetting Ltd,
Bury St Edmunds, Suffolk

Contents

Preface to 1989 Edition

All over the world, the standard of living in different countries, climates and economies primarily depends to an increasing extent on the availability of a reliable supply of electricity, and although the Techniques and Regulations applicable to the distribution and utilisation of electricity vary from country to country, the basic underlying principles and needs remain the same.

Historically too, the story of the use of electricity is one of continual change as new appliances and inventions are created requiring frequent revision and upgrading of existing electrical installations to meet new demands. These changes are reflected in the 15th Edition of the Regulations for Electrical Installations (better known as the Wiring Regulations) which introduced some new approaches to the requirements of previous editions and a new basis for ensuring safety from electric shock.

The general arrangement of the Regulations is quite different from that of previous editions and should be studied carefully by those intent on specialising in electrical installations.

In the United Kingdom, standardisation of the electrical in-stallation regulations with those of the European Economic Community are proceeding and the latest situation is shown in the 15th Edition of the Regulations and the Amendments to them which are issued from time to time. Amendments are also caused by modernisation of designs and improved methods of construction of electrical appliances.

Increasing electrification in the house and home has made it necessary to expand the explanation of domestic appliances, includ-ing workshop and garden tools; developments in space heating, house insulation, electricity in caravans, telecommunications, personal computers and fire detection also need to be included. Today electricity serves so many different purposes that the young electrical technician and engineer in the utilisation field needs to be familiar with the various techniques of several specialist services which he may have to provide for in the house, so the subjects covered by this book have been widened as much as possible.

These, then, are the circumstances which brought about this revised edition. It is intended to give the reader a basic understanding of the principles involved in achieving an adequate and safe supply of electricity to a house which must be in conformity with any local Regulations that may be applicable, and in selecting and using the wiring and equipment that will be installed. Some historical notes are included for information on those locations where a public utility supply is not available. It is hoped that the up-dated information it contains will continue to be useful to all those who are interested in electrical services in the home, but any reader who intends to follow an electrical career must realise that, although the contents of the book will give the layman an insight into the Wiring Regulations, to the extent that his house installation design is based on them, the Regulations as a whole have a much wider scope and cover much more ground and detail than is possible or necessary to include here.

For houseowners it should also be realised that there is no current legislation that prohibits electrical work in domestic premises being carried out by a D.I.Y. enthusiast or indeed any other unqualified amateur electrician.

Care should therefore be taken to ensure that any work carried out is tested by an expert before it is energised and particularly that any electrical work carried out as part of another specialist installation such as a building extension, sun lounge, kitchen, bathroom or shower refit etc., is done correctly and competently.

Due to the lack of controlling legislation a number of unskilled operatives seek electrical installation work and have become a threat to safety by doing jobs that 'seem to work' but are in fact dangerous if not done properly. If in doubt the householder can always refer to the Local Electricity Board or a Registered Member of The Electrical Contractors' Association who will carry out the work required against an agreed quotation or who will arrange for the installation to be tested for a fixed fee.

This book should enable a technically-minded householder to inspect work that has been carried out to decide whether further professional advice should be sought.

Acknowledgement is made of the valuable help of manufacturers named below the illustrations, The Institution of Electrical Engineers, The Electricity Council and others who have contributed comments on the various subjects covered, to whom our sincere thanks are due.

G. Davidson
L. C. Lamb
March 1989

1

Electricity, its Supply, Effects and Measurement

Public electricity supply and regulations

In this country a good supply of electricity is one of the public services we take for granted and, with the rapid development of nuclear power and prospect of cheaper supplies, no house is complete without an adequate electrical installation designed to serve all purposes. As a labour-saving agent electricity is clean, convenient and easily controlled, but it is essential that electrical installations should be fitted by competent electricians, and such important work should not be given to any odd handyman who can use a pair of pliers and a bit of flexible wire. Safety requirements are becoming more complex and stricter, so the competence of contractors and electricians is becoming increasingly important. The National Inspection Council for Electrical Installation Contracting has been in operation for a number of years and inclusion on its Roll of Approved Contractors is regarded as a guarantee of good workmanship; membership of the Electrical Contractors Association is also an indication of good repute.

The supply of electrical energy is the responsibility of a national undertaking with twelve Area Boards, the Central Electricity Generating Board, and an advisory body, the Electricity Council. The Government Departments concerned are the Department of the Environment and the Secretary of State for Energy, which administer the Electricity Supply Regulations 1937. Electrical installations are, in general, designed to conform to the Wiring Regulations of the Institution of Electrical Engineers, which, though not mandatory, are the recognised basis of good wiring practice and are usually specified as an essential part of installation contracts. Statutory Regulations of a more specific nature cover various industrial installations, but the only statutory requirements for domestic equipment are contained in the Electrical Equipment (Safety) Regulations, 1975, the Electrical Equipment (Safety) (Amendment) Regulations, 1976, and the Toys (Safety) Regulations, 1974, administered by the Department of Trade. The Electricity Boards are, however, authorised to refuse to connect or may disconnect from the supply any installations which do not meet with their minimum requirements for safety, which are generally based on the I.E.E. Wiring Regulations. Each area has a Consultative Council which approves tariffs and can deal with consumer problems and complaints not solved by lower level authority. There are also two Area Boards in Scotland.

In other countries, there may not be a National or State Authority which has the responsibility for the supply of electricity to domestic consumers in a house, but this service may be provided by one or more private electricity generating companies each with its own Regulations and Tariffs and offering a degree of competition and choice in source of supply to the consumer.

It may be that current proposals to de-nationalise the C.E.G.B. and Area Electricity Boards in the U.K. will be carried out and that private electricity supply networks will be set up and operated by various companies and individuals as appropriate at the time, but it is likely that important Regulations for safety will not be changed and will be maintained at National level.

Electricity supplies
In remote areas where a public utility supply is not available for the house, some means of generating and storing electricity can be provided.

There are many ways of generating electricity, some being dependent on ample natural supplies of water, fuel supplies of wood, peat, coal, oil, natural gas or geothermal hot water, wind and sunlight. There are also refined fuels such as diesel oil, petrol and paraffin.

Electricity can therefore be provided in all parts of the world by careful selection of an appropriate fuel and the associated generating plant. The capital plant cost and the running costs of fuel and maintenance must also be considered and the generating system engineered for minimum overall cost in operating the plant in a specific location.

The simple private electricity generating plant of a single petrol or diesel driven unit sized to meet the maximum demand that can be anticipated generally results in the minimum initial capital outlay but has a relatively high running cost and poor reliability. The generator runs continuously and for long periods at low load, which are both very bad for maintenance and fuel consumption; therefore a stand-by battery is essential for satisfactory supplies.

It can be shown that by analysing the individual electrical loads, arranging large consumptions such as pumping, refrigeration and washing machines to operate only when the generator is running, dispensing with all electric heating in favour of other natural or direct fuel sources, and by changing low efficiency tungsten lighting units to high efficiency fluorescent sources the minimum loading for the generating plant can be found and it is possible to operate for long periods from batteries charged by the generator. This technique may be supplemented by solar, wind, or other free natural sources also to charge the batteries and thus reduce the peak load on the generator and reduce its running time.

All other matters having a bearing on the period of operation of the generating plant also need careful consideration and these may include maximum use of daylight hours for work periods when the power load is high and battery use at night, giving relief from generator noise when the load is low and during sleep periods. Insulation against heat loss is also extremely important and cost effective, as from the Arctic to the Outback, money spent on house and refrigeration insulation usually shows a good return and modern developments such as cold water detergents and manmade 'drip-dry' fibre clothing, should not be overlooked as they effect large savings by not needing hot water for washing nor electricity for ironing and drying.

Waste heat recovery should also be considered when private generation is necessary as simple heat exchanger or more expensive reverse cycle heat pump systems (see Chapter 9) can dramatically reduce the level of fuel consumption overall and also affect the total running hours of the generating plant. Even the time and cost of obtaining replacement fuel supplies should be taken into account as an item that is often overlooked in remote locations.

Recent studies in various parts of the world having ample wind or solar power, have shown that by careful analysis of the existing situation, the installation of a secondary generating system utilising the natural resource, batteries and control equipment has resulted in a 50% reduction overall in generator size and associated capital and operating costs with a substantial increase in reliability of electricity supplies for essential services such as lighting and communication.

Effects of an electric current

While this book is not intended to be a technical treatise on electricity, some of the elementary effects of electric currents are given in the following sections to help the layman to understand the operation of common electrical equipment.

Electronic effect
An electric current is the flow of electrons through a conductor, the electron being the elementary basis of quantity of electricity and an essential constituent of every atom in all matter. Electronics is the science of the behaviour of electrons and the use of methods of controlling their activity. Whereas 'power' electricity deals with voltages and currents of appreciable magnitude, electronics deals with infinitesimal quantities of electrons the movements of which are called charges rather than currents. Electrons move in all matter, and more easily in semi-conductors than in insulators and differential movements of electrons in the atoms of semi-conductors, such as germanium and silicon, is caused by subjecting crystals to various electrical stresses.

The transistor is a form of junction, in a semi-conductor, of positive and negative regions in the material which can act in a single crystal as conductor, insulator, rectifier or amplifier according to the way it is manufactured and connected in an electrical circuit. The junction has a positive or negative bias (hence the terms P or N junction) when a voltage is applied to it; the purity of the crystal or degree of impurity incorporated in its manufacture regulates the type of junction it is to be used for (P or N), and its resistivity, capacitance and form are controlled in manufacture to suit its purpose in the electrical cricuit. Connections are somewhat similar to those of a thermionic valve. But so many variations of form, construction and connection are possible with transistors that they can be used as oscillators, generators, fast switches and delaying action controls with computer memories, incorporated in the electrical circuitry of the equipment.

The silicon chip is the name given to a miniature integrated (complete) circuit formed on a silicon wafer, which may be self-contained or attached to other wafers of the same or different material, each having a pattern of metallic circuit connections formed by deposition and etching and having joints with others in complex circuits. The circuit components, such as resistors, transistors, capacitors, etc., are all formed on the wafers by deposition, and fine wires connect to the package terminal leads so that the whole package forms a complete unit, sometimes called a microprocesser, and so miniaturised that it may be no larger than the head of a tack or nail, designed for a specific purpose such as a hearing aid or pocket calculator; or to work with many other such chips in more elaborate equipment for industrial process control or large business computers and telephone exchange switching. When applied to a washing machine or other small powered appliances the control circuit may have transistors designed with adequate conductivity for the power current or to control contactors or circuit-breakers which carry the power current of large equipment.

Heating effect
When an electric current flows in a conductor, heat is produced. The electric fire is a common example, in which the element is raised to a red heat. The electric lamp depends on this effect, as the filament is so hot that it is incandescent and emits light. These are useful applications, but if a wire carries too great a current for its size, then dangerous overheating will result, with a consequent fire risk, which must be guarded against. But this effect is made use of in the protective circuit fuse which melts if the current in the circuit is too great for the wiring and equipment to carry safely.

Magnetic effect
Everyone is familiar with the horseshoe magnet, which will pick up iron nails, and other small pieces of iron and steel. The magnetic compass (see Fig. 1) consists of a magnetised needle which is mounted on a pivot and points in the direction of the magnetic meridian. When a wire is carrying an electric current, it is sur-rounded by a magnetic field and behaves in a similar way to a magnet. If the wire is wound up into a coil, a very much stronger effect is produced, which can be made greater still by winding the wire on an iron core. But this large magnetic effect is only present while the current is flowing, so this property is employed to obtain a magnetic field when required, and the combination is called an 'electromagnet'. Special hard steel can be given permanent magnet-ism by such a method, i.e. it remains magnetised when the electric current in the coil is discontinued. On the other hand, soft iron loses practically all its magnetism when the current ceases. In the con-struction of electromagnets, soft-iron cores in the form of solid rod, iron wire or strips of thin iron sheet are employed. This magnetic effect is employed in electric bells, telephones, clocks and indicating devices. If a wire carrying a steady electric current is placed over a magnetic compass needle, as in Fig. 1(b), then with the current flowing in one direction the needle will be deflected one way, and on reversal of the current the direction of movement is reversed. This principle is employed in various types of instruments and meters used for measuring electric pressure, current and power, besides its

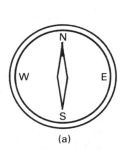

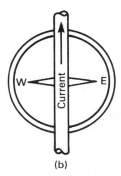

Fig. 1 (a) Magnetic compass pointing north and south

(b) Current in a conductor deflecting a magnetic compass

application for power purposes to motors, generators, transformers and other devices. If electric conductors adjacent to one another carry excessive currents, then, in addition to the resulting dangerous heating effect, there is a magnetic force between them which can cause movement and damage. This does not occur in domestic installations with relatively low currents.

Chemical effects

If two wires from a small dry battery are immersed in a glass containing water, to which a few drops of vinegar are added, bubbles of gas will be seen to form on the wires when they are brought sufficiently close together *without* touching. This shows that the water has been split up into its constituent elements, oxygen and hydrogen, which evaporate as gases. If the same wires are stuck in a potato, about 2 mm apart, then the wire connected to the positive terminal of the battery will show a green discoloration due to the effect of the current on the chemical constituents of the potato. *Do not try this with the mains supply as it is very dangerous.* This effect is usefully employed in the electrodeposition of metals and in refining. It is also evident as apparent corrosion when there is a leakage of electric current from a cable or conductor to adjacent metalwork or damp surfaces. This is frequently seen on top of a car battery when electricity leaks across a moist surface between the terminals. The chemical effects are only produced with direct current (d.c.).

Arc effect

If two carbon rods, connected to a suitable supply, are momentarily touched together and then separated, an intense spark called an 'arc' is formed. With sufficient electric pressure across the gap, the current does not cease but persists in the form of an arc. The temperature of the carbon rods is about 3500 degrees Centigrade and an intense white light is emitted, mainly from the incandescent carbons. This effect has been usefully employed for searchlights, photographic processes, electric furnaces and electric arc-welding. An arc is produced in electric fluorescent lamps and other gaseous discharge lamps used in street lighting. A similar effect can occur with a break in an electric cable, when the arc may persist between the broken ends or to any adjacent metal-work, constituting a fire risk. Arcs also occur at switch contacts and wherever the current is interrupted by breaking a circuit in the wrong way instead of using a switch, such as disconnecting a plug and socket carrying direct current or with certain types of inductive circuits using alternating current.

Physiological effect

Electric shock to the various parts of the body is dangerous and may be fatal. For this reason, care should be exercised with all electrical connections. Switches should be 'off' or the fuses withdrawn when any alterations or repairs are being made to the wiring of a house. 'Safety first' must be considered at all times with electrical apparatus. Chapter 13 deals more fully with these aspects.

Electrical units

Before we are able to compare the sizes of different objects we must have some system of suitable units. A tape measure is suitable for measuring a length of cloth, but it is of no use in weighing coal or finding the hours of sunlight in a certain day. The scientific basic units include the *metre* (m) for length, the *kilogram* (kg) for mass and the *second* (s) for time. All the practical electrical units are related to these dimensions, but we are only concerned with those units in everyday use.

Electricity is a form of energy; and energy is defined as the ability to do work. So, when energy is expended, the work is done. Electrical energy is charged for as units in the electricity bill, and the account may include a fixed charge and some charges for other services, such as the hire-purchase of an electric cooker.

The legal unit for the sale of electrical energy is the *kilowatt-hour* (abbreviated to kWh) and is usually referred to simply as the 'unit' of electricity.

Energy is equal to the rate of doing work multiplied by time, and the rate of doing work is called *power*. In everyday speech a motorvan may be said to be more 'powerful' for shifting a load than a horse and cart, meaning that the former does more work in a given time than the latter. If a labourer carried 10 kg of sand 10 m up a ladder, then 981 joules of energy have been expended, whether he takes 2 minutes or 2 hours over it; but in the former case the rate of doing work is sixty times faster than in the latter, and the power is therefore sixty times greater.

The electrical unit of power is the *watt* (symbol W). It is used when we say that a 60-watt lamp is more powerful than a 25-watt lamp. The unit for a quantity of electricity is the *coulomb* (symbol C), analogous to the litre as a quantity of water; but we are only concerned with the rate of flow of electricity, or current. This is the quantity of electricity flowing in a circuit in a unit of time, the second, and it is called the *ampère*, often abbreviated to amp or A. In a water main we are not concerned with how many litres it will hold but with the delivery in litres per second; so, in just the same way, we are interested only in the flow of current in an electric cable or how much current we must make flow in an electric fire to make it red-hot. A level water main full of water would not deliver any water to the domestic kitchen tap without pressure; thus pressure affects the *rate* of flow. This is obtained either by the height of a reservoir above the main or by the pumps at the waterworks. In the electricity main or cable it is the pressure generated at the power station that causes the electric current to circulate around the supply system. This pressure is measured in *volts* (abbreviation V) and is present all the time, whether we have the electricity turned on, by means of a switch, or not. Thus the electrical power available depends on both the pressure and the intensity of flow. Hence *electrical power in watts is the product of the pressure in volts and the current in ampères*, or, with letter symbols, $W = V \times I$.

The difference in pressure, or *potential difference* (abbreviated to p.d.), across the ends of a circuit causes a current to flow if the circuit is complete. But what settles the magnitude of the current? If we have a new, clean water main, very little pressure will send the water along it, as the friction between the moving water and the walls of the pipe is very low. Suppose the main is half full of sand and gravel, then a very much higher pressure will be needed to deliver the same quantity of water in a given time; also, a smaller clean pipe would require a greater pressure and velocity to deliver water at the same rate. The total resistance to flow will depend on the length of the pipe, its internal cross-sectional area and its material wetted surface. In a similar manner, the conductor material, such as copper, of an electrical circuit is like the bore of a water main and offers a resistance to the rate at which electricity flows through it. So it can be said that the rate of flow of electricity increases with the pressure and decreases with the resistance. This statement is the basis of Ohm's Law, which can be stated as: *Current is proportional to the voltage and inversely proportional to the resistance.*

The unit of resistance is the ohm (symbol Ω). A p.d. of 1 volt applied to a resistance of 1 ohm will cause a current of 1 ampère to flow; alternatively, volts divided by ohms gives ampères. With symbols, I stands for current, V for voltage and R for resistance; then:

$$I = \frac{V}{R}, \text{or} \quad V = I \times R, \quad \text{or} \quad R = \frac{V}{I}.$$

If any two of these quantities are known, the third one can be found.

Power, in watts, is the product of voltage and current, therefore:

$$P = V \times I.$$

This product is true power only in direct current circuits and has to be multiplied by another quantity, called the 'power factor', to give true power in alternating current circuits because certain a.c. equipment characteristics cause the voltage and current alternations to be out of phase (not coincident in time) with each other. But in household installations consideration of this refinement is generally unnecessary. The watt is too small a unit for many installations, so the term *kilowatt* is employed, which is 1000 watts, and 1 kilowatt for 1 hour is 1 unit of energy (1 kWh).

EXAMPLE 1.   An electric fire, having a resistance of 40 Ω, is connected to a 200 V circuit. What current will it take, and what will be the power in kilowatts? If the cost of power is 3p per unit, how much does it cost to use this fire for 3 hours?

$$\text{Current in amps} = \frac{\text{volts}}{\text{ohms}} = \frac{200}{40} = 5\,\text{A}.$$

$$\text{Power in watts} = \text{volts} \times \text{amps} = 200 \times 5 = 1000\,\text{W}$$
$$= 1\,\text{kW}.$$

$$\text{Energy} = \text{Power} \times \text{Time} = 1\,\text{kW} \times 3\,\text{h} = 3\,\text{kWh}.$$

If 1 kWh is 1 unit costing 3p, then cost = $3 \times 3\text{p} = 9\text{p}.$

The old mechanical unit of power was the horsepower, which is equivalent to 746 watts. Under the metric system motor ratings are stated in watts for machines that are driven by electric motors.

In the case of alternating current (a.c.) circuits and a.c. apparatus windings, the values of current and voltage are not coincident (in phase) due to inductive effects, which make the current lag behind the voltage. Capacity effects make the current lead the voltage. Therefore, the product of current and voltage values will not give the true power. In a.c. calculations this product is called *volt-ampères*, not watts, and the two different values of voltampères and true power watts are related by the *power factor*, which is the ratio

$$\frac{\text{Watts}}{\text{Voltampères}} \quad \text{or} \quad \frac{\text{W}}{\text{VI}}$$

With a load of pure resistance the power factor is unity, or it may be expressed as 100%; but with inductive apparatus it is less than one, or below 100%. Low power factor means that more current is required for a given power, and as this adversely affects the supply system it is often penalised in tariffs.

The combined effect of resistance and inductance of an a.c. circuit is called impedance, symbol Z, and is used instead of resistance, R, in the simple Ohm's Law equation given on page 9.

Current required by motors

The output at the motor shaft is given in watts. Due to the motor losses, or power absorbed by the working parts in friction, windage and resistance, and converted to heat, the watts input at the motor terminals must be greater than the output, and the ratio of these two quantities is a measure of the *efficiency*.

$$\text{For any machine, Efficiency} = \frac{\text{Work output}}{\text{Work input}}.$$

$$\text{For a d.c. motor, Efficiency} = \frac{\text{Watts output}}{\text{Volts} \times \text{Current input}}.$$

$$\text{and Current input, in ampères} = \frac{\text{Watts}}{\text{Volts} \times \text{Efficiency}}.$$

For small d.c. motors with an efficiency of about 75%, the current required from the supply is given by:

$$\frac{\text{Watts} \times 100}{\text{Volts} \times 75}.$$

With a.c. motors, the power factor as well as losses still further increase the input current. With small motors, the power factor varies from under 70% to over 80%. Thus, for an a.c. motor:

Current input, in ampères

$$= \frac{\text{Watts output}}{\text{Volts} \times \text{Efficiency} \times \text{Power-factor}}.$$

Taking an average value of 75% for both efficiency and power factor, the current taken from the supply is given by:

$$\frac{\text{Watts} \times 100 \times 100}{\text{Volts} \times 75 \times 75}, \text{ or approx. } \frac{1 \cdot 8 \text{ W}}{\text{V}}.$$

Measuring instruments

The pressure is measured by a voltmeter, which is *connected across* the mains or the apparatus concerned (in parallel). This is like a pressure gauge on a boiler; this does not measure the steam consumption and in a similar way, the voltage is not a measure of the current used.

The current is measured by an ampèremeter, usually called an ammeter, which is *connected in* the circuit, like a gas meter in the gas main (in series).

The correct method of connecting a voltmeter and an ammeter is shown in Fig. 2.

It is possible to connect the voltmeter across the supply mains, as it has a high resistance and thus will not take much current.

Modern instruments commonly used for checking alternating current (a.c.) are the Megger instruments shown in Figs. 3 and 4 which measure voltage by using test leads which are plugged into one end of the instrument. The jaws enclose a single core cable, and the current is measured by induction or transformer action and

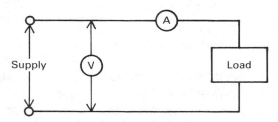

Fig. 2   Correct method of connecting a voltmeter and an ammeter

Fig. 3   General purpose Avometer* (arranged for clip-on use)
(*Megger Instruments*)
* Registered trademark

indicated on the selected scale; this is not possible with direct current (d.c.), see Fig. 3.

EXAMPLE 2.   A voltmeter that will read up to 250 V has a resistance of 10,000 Ω. What current will it take when connected across a 200 V d.c. supply?

$$\text{Current in amps} = \frac{\text{volts}}{\text{ohms}} = \frac{200}{10,000} = \frac{1}{50}, \text{ or } 0.02 \text{ A}.$$

The ammeter, on the other hand, must *never* be connected across the supply, whether it is from the mains or a battery, as it would take a very large current, due to its low resistance, and suffer internal damage.

EXAMPLE 3.   An ammeter that reads up to 20 A has a resistance of $\frac{1}{100}$ Ω. What current will it take if accidentally connected across (a) a 12 V car battery and (b) the 230 V mains supply? Theoretical values, neglecting battery resistance, would be:

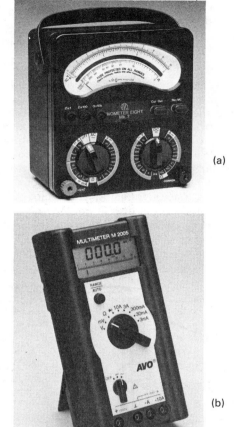

Fig. 4 (a) Analogue (dial) Avometer, **(b)**, Avo* analogue/digital (numerical) multimeter (*Megger Instruments*)
   * Registered trademark

(a) Current $= \dfrac{12}{0.01} = 12 \times 100 = 1200$ A.

(b) Current $= \dfrac{230}{0.01} = 230 \times 100 = 23{,}000$ A.

In the latter case the circuit fuses or circuit-breaker would cut off the supply, but the ammeter would be damaged in either case and the effects would be extremely dangerous.

EXAMPLE 4.   What voltage is required across the same ammeter to give the full deflection of 20 A?

By Ohm's Law, Volts = Current × Resistance

$$= 20 \times \frac{1}{100} = \frac{1}{5}; \text{ or } 0.2 \text{ V}.$$

On d.c. systems the product of voltage and current can be measured by a wattmeter, but an ammeter and voltmeter can be used to serve the same purpose. An a.c. wattmeter is a similar instrument but with the movement designed for alternating current.

Fig. 5(a) illustrates a type of multimeter testing set in common use. These have a selection of several ranges of volts and ampères, including milliamps, as well as resistance, measured in ohms, for d.c. and a.c. measurements. Some multi-meter instruments also include scales of capacity, induction and decibels and are also made with a digital liquid crystal display reading, as shown in Fig. 5(b) instead of a moving coil movement with pointers and scales. These types of instrument are very compact and depend on electronic internal circuitry for operation with small dry cells for current supply.

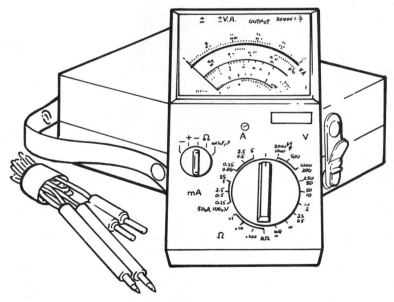

Fig. 5(a) Portable d.c./a.c. multimeter testing set
(*Crompton Instruments*)

Fig. 5(b) Hand-held multimeter with digital reading
(*Robin Electronics Ltd.*)

Supply meters

These meters are the property of the supply undertaking and are sealed on installation so that they cannot be tampered with by unauthorised persons. They combine the measurement of volts, ampères and time, and depend for their action on the magnetic effect, although on some direct current systems an electrolytic type which utilises the chemical effect of a current has been employed in the past. Instead of a pointer, as in an indicating instrument, the common single-phase meter has a moving element that rotates and drives a train of wheels, which records the number of units consumed (kW hrs) on a number of small dials. The rotating disc runs between the poles of an electromagnet, and the speed of the disc

(a)

(b)

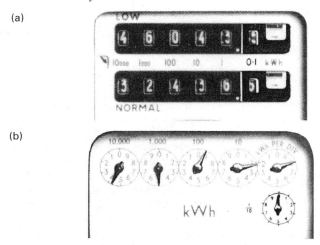

Fig. 6   Electricity meter dials
(a) Economy 7 digital meter reading 46043 low rate and 32436 normal rate
(b) Dial meter reading 44928

varies with the current passing through the magnet coils due to eddy currents induced in the disc and producing a motor effect.

How to read an electricity meter
The reading on any meter is the continuous summation of all units used since it started at zero, therefore the consumption of electricity for a particular period is only obtained by subtracting the reading at the beginning of a given period from the reading at the time of inspection at the end of that period. The digital meter Fig. 6(a) gives a direct reading of the number of units used.

A dial meter is not quite so easy to read and gives the consumption of electricity by pointing to numbers on clock. On the older, smaller sizes there are four black dials which register thousands, hundreds, tens and single units. The hands of adjacent dials revolve in opposite directions. The red dials, which register in tenths and hundredths of a unit, may be disregarded since they are provided for testing purposes. Typical meter dials are shown in Fig. 6(b).

To read the dial meter start at the right-hand dial of single units. When the hand is between two figures, write down the *lower* figure; if between 0 and 9, always write down 9. Repeat the process with the other dials, writing down the figures in the order right to left. If the hand is *on* a figure (say 6), write down 5, not 6, unless the hand on the previous right-hand is between 0 and 1. On larger meters there are five black dials going up to ten-thousands.

Charges for electricity

There have been many scales of charges in different parts of the country in the past and some perplexity has been caused by the variety of methods employed. It was realised in the Industry that rationalisation was desirable, and considerable progress towards standardisation of tariffs has been made by the Area Boards.

Unfortunately, electricity cannot be generated at a steady rate and stored during the periods of low demand, like gas, except for the relatively small amount that can be stored in secondary cell batteries. The power stations have to be able to supply whatever amount is called for 'on demand'. This means that the generating plant installed must be large enough to supply the 'peak' load, not just the average load on the system. The National Grid enables fewer and larger power stations to be operated more efficiently by spreading peak loads over a number of interconnected stations, and maintains a more reliable supply than many independent and separate local stations could. Typical daily load curves are shown in Fig. 7 and it should be noted that the winter load is of similar shape to, but almost twice the height of, the summer load; and that the Sunday load is also similar in shape to, though lower than, the weekday load, except for a greater dip in the afternoon between 1300 and 2000 hrs. These curves show the combined effects of industrial and domestic demands and also illustrate why off-peak loads are encouraged by the Electricity Boards – to fill the valleys in the curves and so obtain more efficient use of generators.

The capital charges on the plant, together with the other standing charges, must be met, as well as the running costs of generation and distribution losses, which vary with the load.

Based upon typical tariffs as issued by one Area Electricity Board the standard charges for electricity where the maximum power required does not exceed 40 kVA are:

1. D1 DOMESTIC GENERAL TARIFF. A single standard unit price is charged per kilowatt/hour for all electricity used with a small 'fixed' or standing charge per week which is increased if a prepayment meter (coin, key or slot) is fitted.
2. D2 DOMESTIC ECONOMY 7 TARIFF. This is the ideal tariff for the average small family household with electricity used for hot

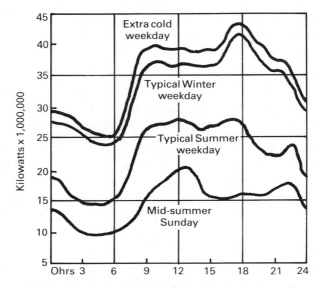

Fig. 7   Typical load curves (*The Electricity Council*)

water storage at night and with night storage heaters for daytime space heating, with or without a midday boost.

The period described as 'Night' is identified as any seven hours taken between 10 p.m. and 8 a.m. (with adjustment for British Summer Time); hence it is known as the 'Economy 7' tariff.

The basis of the tariff is that all electricity consumed during the seven hours is charged at less than half the cost of the D1 tariff and is recorded on a separate meter or dials for account purposes. Electricity consumed at all other times of the day (or night) is also separately metered and charged at the standard D1 rate. It is therefore advantageous to delay all major electricity usages such as water heating, space heating, clothes and dish washing etc. to coincide with the seven hours at cheap rate.

It should be noted however that the fixed or standing charge is increased where the Economy 7 tariff is adopted by an amount typically representing about 100 units per quarter at the difference between D1 standard rate and the Economy 7 rate. Therefore unless a minimum of 100 units are consumed per quarter at the Economy 7 rate there is no overall saving.

3. D3 DOMESTIC NIGHT AND DAY TARIFF.   This tariff was designed to enable consumers to take advantage of off-peak night rates between 9 p.m. and 7 a.m. (which are higher than the D2 Economy 7 rates) without having to install separate installations for off-peak metering but it is not now generally available except to existing users. Two meters were often fitted, or a meter with two sets of dials, and by means of a separate time switch, one meter registers the night consumption and the other registers the day consumption. The two quantities are charged at the appropriate rates and in addition there are weekly fixed or standing charges. There may also be another off-peak period during the day at the cheap rate and covered by the same time clock and metering.

SMALL BUSINESS TARIFFS.   These tariffs are also only available where the maximum power requirement does not exceed 40 kVA and are intended for combined private and business premises, small farms, small shops, etc.

4. BE1 SMALL BUSINESS GENERAL TARIFF.   This is similar to the D1 tariff but with a higher price per unit for the first block of 1001 units per quarter (or in the case of combined business and domestic premises, 494 units per quarter) and the same rate as D1 for the remainder of units used. The fixed or standing charge is the same as for the D1 tariff.

5. BE2 SMALL BUSINESS ECONOMY 7 TARIFF.   This is similar to the D2 tariff but with a higher price per unit for the first block of 1001 or 494 units per quarter taken other than during the 'Night' period of any 7 hours between 10 p.m. and 8 a.m.

6. BE3 SMALL BUSINESS EVENING WEEKEND ECONOMY 7 TARIFF.   This is similar to the BE2 tariff but with a lower standing charge and an increased charge for the blocks of 1001 or 494 units and all other units taken in Daytime (7 a.m. to 7 p.m.) from Monday to Friday with a much reduced rate for Saturday/Sunday and times outside the stated periods.

There are other tariffs of limited availability for application to consumers with existing tariffs that do not correspond with the latest standard tariffs as well as Restricted Hours Tariffs with low standing charges and unit rates for electricity supplies taken at various times that correspond to low generating demand periods of early morning, afternoon and at night.

There are also several Maximum Demand Tariffs for supplies in excess of 40 kVA such as factories, large offices, farms and process plants and Seasonal Time of Day Tariffs, all at attractive rates and intended to help even out the load curve shown in Fig. 7.

In summarising the typical tariffs it can be seen that the flat rate D1 is more suitable for very small flats or houses with no electric heating for space or water and with a consumption overnight of not more than 100 units per quarter.

With additional heating and cooking loads or where a business or profession is combined with the normal occupation of the house, it may be more economical for the householder to use the tariffs D2, D3, BE1 or BE2, thus ensuring a reducing average cost per unit with higher consumption.

The Night and Day or Economy 7 tariffs favour the off-peak use of electricity for storage space and water heaters, or any purposes such as automatic clothes or dishwashing during off-peak hours at low rates. A separate off-peak tariff requires separate meters or dials, time switches, controls and a wiring installation separated from the normal installation so that it cannot be used during the daytime peak hours and was originally intended for storage heating installations only.

EXAMPLE 5.   A householder has the choice of paying for electrical energy either:

(a) at a flat rate of 6p per unit and 60p per week standing charge,

or   (b) at block rates of 300 units at 10p per unit, 1000 units at 8p per unit and 4p per unit for the remainder.

Which would be the cheaper tariff if the average quarterly consumption is

(i) 1,000 units, or (ii) 5,000 units?

At what number of units would tariff (b) be more economical?

(i) 1000 units per quarter would cost:
  Scale (a) = (1000 units × 6p) + (13 weeks × 60p)   = £67·80
  Scale (b) = (300 units × 10p) + (700 units × 8p)   = £86·00
(ii) 5000 units per quarter would cost:
  Scale (a) = (5000 units × 6p) + (13 weeks × 60p)   = £307·80
  Scale (b) = (300 units × 10p) + (1000 units × 8p)
                     + (3700 units × 4p)   = £258·00

Scale (a) is cheaper for 1000 units while scale (b) is cheaper for 5000 units per quarter.

The answer to the latter part of the question can be arrived at by calculation to find the point at which the tariff curves cross over, i.e. they are equal at 'x' number of units:

$$(x \text{ units} \times 6p) + (13 \text{ weeks} \times 60p) = (300 \text{ units} \times 10p)$$
$$+ (1000 \text{ units} \times 8p)$$
$$+ ([x - 1300] \text{ units} \times 4p)$$

which when simplified:

$$6x + 780 = 3000 + 8000 + 4x - 5200$$
$$6x - 4x = 3000 + 8000 - 5200 - 780$$
$$2x = 5020$$
$$x = 2510 \text{ units.}$$

Alternatively, the answer is most easily illustrated by a graph: by calculation using the above tariffs for 0 to 5000 units a number of points can be found which can be plotted to form the graph given in Fig. 8.

Scale (a)

Units per quarter	0	1000	2000	3000	4000	5000
Total cost (£)	7·8	67·80	127·80	187·80	247·80	307·80

The above values are shown on the graph Fig. 8 as points and are joined by a straight line titled 'Flat Rate Tariff'.

Scale (b)

Units per quarter	0	300	1000	1300	2000	3000	4000	5000
Total cost (£)	0	30·00	86·00	110·00	138·00	178·00	218·00	258·00

The above values are shown on the graph Fig. 8 as triangles and are joined by a curve titled 'Block Rate Tariff'.

A sufficient number of values should be taken to straddle the estimated consumption of the premises concerned. Where these two graphs cross one another shows the number of units per quarter at which both tariffs are equal in cost and, on either side, which tariff offers lowest cost for any given consumption. These graphs are intended to show the advantage of block rates, as the electrical consumption increases; with the use of more domestic and other electrical appliances, the average cost per unit falls.

Curves for other forms of tariffs can be plotted in the same way, the only difference being that the flat rate tariff curve will have a higher standing charge and may slope less than the block tariff curve with a lower unit price, while a curve for day and night tariff will be somewhat similar but the slope will be less with greater off-peak consumption.

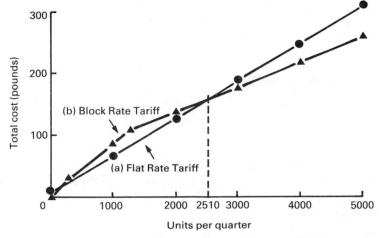

Fig. 8   Graphical comparison of electricity charges

All these tariffs generally have specified prices of units which may be liable to vary from time to time with variation in the cost of fuel used in the power station.

It is also worth noting that where landlords of flats and rooms charge tenants for metered electricity used, it is illegal to charge more per unit than the figure prescribed by the Electricity Boards for such a purpose.

The data for the examples is, of course, slightly simplified for clarity and the exact tariff figures will have to be obtained from the Electricity Supply Authority in the area in which the house is situated.

2

Conductors, Insulators and Circuits

Electrical conductors

To transmit electrical energy to the consumer, cables of various types and sizes are employed consisting of central conductors, analogous to the bore of a water main, and the surrounding insulation, which can be likened to the wall of a pipe (see Fig. 9). Any material that allows an easy passage to electricity is called a conductor. All metals are conductors of electricity, but some are better than others. Conductors are required to provide an easy path for the current and should not have a high resistance, otherwise both drop of pressure and power loss will occur which will cause unwanted heating. Low-resistance conductors are therefore required to carry current to the points of utilisation and control, such as the wiring connecting the switches and lamps to the supply. The most common metal for wires and cables is copper, which is specially refined to have high conductivity. Copper is easily tinned and soldered, and is also a ductile metal.

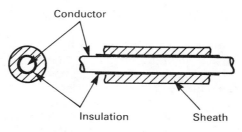

Fig. 9   Single-core cable

Aluminium is also used, particularly for large sections, to carry heavy currents, but special solders and flux are necessary for jointing. It is used in some house wiring cables but special precautions are necessary at contacts and terminals because a high-resistance oxide forms on the surface in air unless plated or tinned during manufacture.

Brass is employed for contacts and terminals, e.g. the brass plungers inside a lampholder and plug and socket contacts.

Carbon is a non-metallic conductor and is used for sliding contacts on brass or copper, e.g. the carbon brushes that lead the current into the revolving armature of a motor. The lead of a pencil is a conductor, and it is possible to get a shock from a lead pencil if it is poked into an electric socket.

Iron is a poor conductor due to its higher resistivity and is seldom employed; its magnetic properties also cause extra losses.

Factors affecting the resistance of a conductor

The greater the length of a conductor, the higher will be its resistance, and for a given length and material the resistance can be lessened by increasing the area of cross-section. If l is the length of a conductor, a its cross-sectional area and ρ its resistivity, which depends upon the material of the conductor, then resistance is given by

$$R = \frac{\rho \times l}{a} \text{ohms.}$$

The 'resistivity' of any material is the resistance between the opposite faces of a cube of the conductor which has an edge of unit length, the value of resistivity depending on the unit used (see Fig. 10). Since copper has high conductivity it has low resistivity, and the resistance between the opposite faces of the cube is very small. Very small numbers, less than one, can be written as a decimal with a number of noughts in front of the first significant figure, but for convenience we denote thousandths by 'milli' and millionths by 'micro', placed in front of the unit name. Thus 5 milliampères is five thousandths of an ampère, or $\frac{5}{1000}$ ampère, and 2 microhms is two millionths of an ohm, or $\frac{2}{1000000}$ ohm. For copper the resistivity, ρ, is approximately 0·0172 ohm per square millimetre section per metre length. As metals get hotter their resistance increases, though the opposite effect occurs with carbon, but the above value for copper is sufficiently accurate for our purpose.

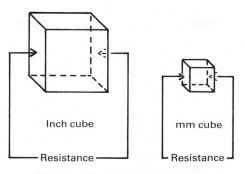

Inch cube

mm cube

Resistance

Resistance

Fig. 10 Illustrating resistivity

EXAMPLE 6. A copper wire has a diameter of 1·78 mm and a cross-sectional area of 2·5 mm². If its resistivity is 0·0172 in metric units (mΩ), what will be the resistance of a length of 1 km?

$$R = \frac{\rho \times l}{a} = \frac{0 \cdot 0172 \times 1000}{2 \cdot 5} = 6 \cdot 9 \; \Omega \; \text{approx.}$$

(Convert the length in kilometres to metres and note that all the units must be of the same kind.) This size of wire is used in cables, and 1 km of single strand will weigh about 22·2 kg without the covering of insulation. As approximate figures for calculating weights, copper weighs about 8·9 g/cm³ or about a third more than iron.

EXAMPLE 7. If the above copper wire of 2·5 mm² were cut in half and run side by side, what would now be the resistance of the two wires connected together in parallel?

The length is now 500 m, but the cross-sectional area is doubled to 5 mm², so by proportion:

$$R = 6 \cdot 9 \times \frac{500}{1000} \times \frac{2 \cdot 5}{5} = 1 \cdot 73 \; \Omega.$$

Some materials have too high a resistivity to be called conductors but they permit electron movement to take place when under electrical stress, so they are called semi-conductors. Their use in electronic circuits in transistors and silicon chips is of the highest importance in modern technology.

Insulators

Electricity will 'escape' to exposed metalwork and earth if not prevented, and it is therefore necessary to insulate the conductor. Insulators separate the conductor from other conductors, exposed metalwork and earth. Insulators are substances that are poor conductors of electricity and serve to prevent, as far as possible, an electric current straying from its conductor path. The pressure or voltage on an electric cable, while sending the current along the conductor, also tends to force a very small current through the wall of insulation around the conductor. If the insulation is insufficient or faulty there will be a leakage current, which will flow to the return conductor (short-circuit) or to earth and return to the power station by any easy path, such as the water mains. Due to the contents of the soil, chemical action will occur, causing corrosion, but this is more likely with direct current than with alternating current systems. The insulation can be likened to the wall of a water pipe, which will only stand a certain hydraulic pressure before it begins to leak and may eventually burst. In a similar manner, if the pressure or voltage on an insulated wire is increased beyond a safe limit, the insulation will break down. Electric wires and cables are insulated for definite working pressures and should not be used for higher voltages. Thus a length of bell wire is amply insulated for the 3 V pressure obtained from two dry batteries, but it is most unsafe to use such wire for connecting an appliance to 240 V mains.

The resistivity of insulators is very high, and instead of writing a very large number with many noughts behind it we denote thousands by 'kilo' and millions by 'mega'. Thus 2000 watts is called 2 kilowatts, and 600,000,000 ohms is called 600 megohms ($M\Omega$). The insulation resistance of cables is measured in megohms and decreases with increase of length, as there is a larger area of insulation – of the same thickness – for the leakage current to pass through.

An example of electronic activity in an insulator is the capacitor, or condenser as if was called originally, in which the electrons in the insulating medium between plates or electrodes of metal are under stress when a voltage is applied to the plates. Thus a charge of electricity is absorbed and allowed to discharge when permitted by circuit control.

Examples of insulators are india-rubber, both pure and vulcanised, polyvinyl chloride compound (p.v.c.), guttapercha, ebonite, mica, glass and slate. Some insulators absorb moisture, such as paper, cotton, silk, various artificial fibrous materials, magnesium oxide and asbestos. As a general guide, good heat insulators are also electrical insulators, so that they tend to retain any heat in the conductor generated by the passage of the electric current. Continual overheating will cause some insulators, such as rubber compounds, to become hard and brittle, and will soften others, such as p.v.c., so cool operation is essential for safety.

Simple circuits

The two simple arrangements of conductors and connected apparatus in a circuit are called 'series' and 'parallel'. The six cells in a 12 V car battery are connected in series as shown in Fig. 11. Since each cell gives 2 V, the total pressure adds up to 12 V, and the current is common through each cell. If 8 A are taken for lamps, then this amount of current flows through each cell. Two dry cells for an electric bell each give 1·5 V. In Fig. 12 they are shown connected in series at (a), when the total voltage is 3. In (b) the similar terminals are connected together, when the pressure is only that of one cell, namely 1·5 V; but each cell will supply half the current, i.e. the total current in a parallel circuit is the sum of all the currents.

With batteries there must be a positive and negative terminal. The series and parallel connections are shown diagrammatically for a four-cell battery in Fig. 13(a) and (b) respectively.

Electrical energy from the mains is distributed with parallel connections, for which the voltage is maintained constant, and all the separate currents add up to give the total load current on the

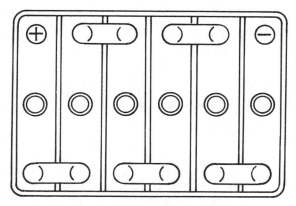

Fig. 11   Car battery, six cells in series

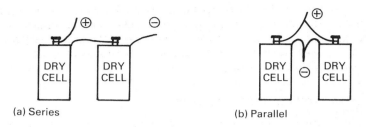

Fig. 12   Battery connections

system. The same parallel method is employed inside the house, the various currents adding up to the total current supplied. But the outlet-point wiring, switch and apparatus are connected in series, since the switch must open and close the circuit, and all these items will carry the same current.

EXAMPLE 8.   Suppose that in one room of a house an electric iron is being used; the iron will take 2·1 A, and a 60 W lamp will require 0·25 A. In another room the family is listening to the radio or watching television in front of an electric fire with one bar switched

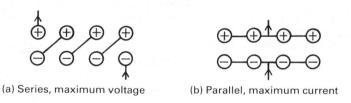

Fig. 13   Series and parallel connections

on. The TV set will take 0·5 A, a 100 W lamp 0·42 A and the 1 kW fire 4·2 A, so the total current supplied to the premises is:

$$2·1 + 0·25 + 0·5 + 0·42 + 4·2 = 7·47 \text{ A},$$

say, 7·5A in round figures.

The size of wire for each piece of apparatus must be large enough to carry the current and must also be properly insulated to withstand the common pressure of 240 V. The switches are in series with the individual appliances they control, as shown in Fig. 14. From this it will be seen that each piece of apparatus has a common voltage of 240 and that the various streams of current combine to give a total of

7·5 A through the electricity meter. Note the switch on the electric fire for the second element; the switch socket will disconnect the fire completely. The radio or TV set will also have its own switch, but the pressure is on the connecting lead as long as the switch of the switch socket is closed.

Fuses and circuit protection

The heating effect of an electric current can become dangerous, but this effect can also be used to open a circuit in which the current is excessive. Fuses are provided to protect both the wiring and connected apparatus, and in so doing protect the adjacent structure of the building against fire. Fuses are sometimes called 'cut-outs' because they cut out the defective circuits. The term fuseboard is given to the complete assembly of a case containing a group of fuses and contacts, but the term distribution board is more general, as it also covers groups of automatic switches that disconnect the circuits when overloads occur. These miniature circuit-breakers are increasingly employed in modern installations, but fuses are cheaper and will therefore continue to be used for some time.

Fuses consist of short lengths of a suitable wire that will safely carry the normal current of the circuit, but with any sustained overload the wire melts. A very large current will blow the fuse immediately, while a lesser current may take several minutes to melt the wire.

For simplicity no fuses were included in Fig. 14, but the various lighting and heating circuits would be divided up and suitably protected by fuses. The electric iron would be supplied through a flexible lead from the switch socket, and a fuse would be inserted between the connecting wires and the supply terminals, as shown in Fig. 15. If everything is in order, the current will not cause any heating of the fuses and current will be supplied to the iron. Now, suppose that the two wires of the flexible lead are worn and that so much of the insulation is rubbed off that the two bare wires can come into contact at the point X. This would cause a 'short-circuit', and the current would take a short or easy path of very low resistance at the fault instead of going through the heater element of the iron. The current would surge up to a very large value and the circuit fuse would melt. The current would be automatically cut off before any damage was done to the rest of the circuit, as it could not get across the gap left by the melted fuse.

It is essential that the correct fuse should blow when a fault occurs and that a fuse should not blow unnecessarily. For this reason, the

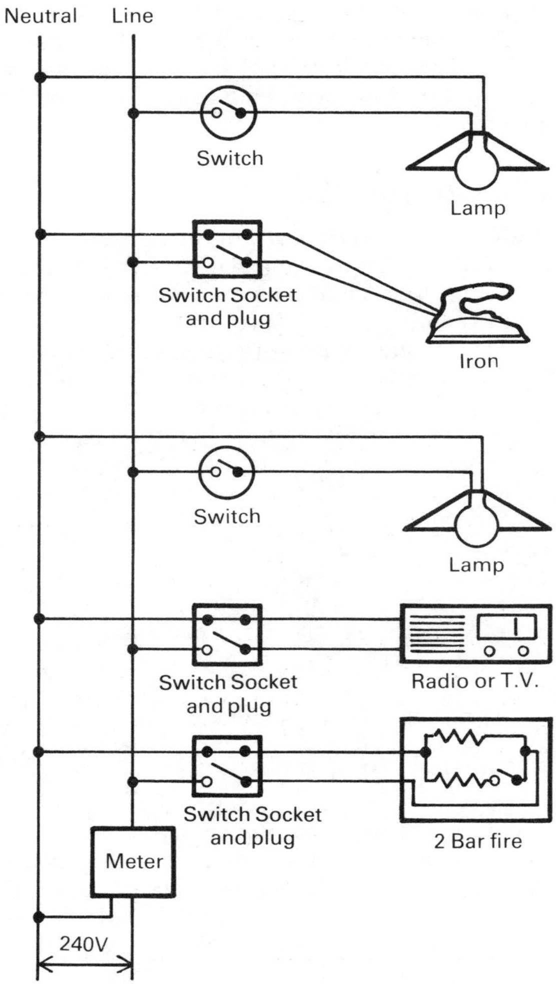

Fig. 14   Parallel connection of apparatus
(protective earthing conductors and fuses not shown)

sizes of fuses are graded to suit the various circuits, such as 5 A for the lighting circuit and 20 A for heating circuits. The main fuses should not melt and cut off the whole supply to a house when a fault occurs on a sub-circuit; this points to bad installation design or

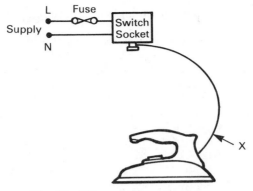

Fig. 15   Effect of short-circuit

wrong choice of fuse. Therefore, it is important to replace 'blown' fuses with the correct size of fuse wire or fuse cartridge, which should have the current-carrying capacity of the smallest cable or flexible cord in the circuit it protects. The sizes of fuse wire for rewireable-type fuses are given in Table 1a.

If a fuse blows repeatedly, it is wrong and dangerous to replace the fusewire with a larger size than it should have or with any odd wire that happens to be at hand. The proper course is to investigate the cause of the fusing and remedy the fault. Cartridge fuses are safer and operate more quickly than rewireable fuses. The miniature circuit-breaker is the best form of circuit protection, and the extra expense is generally worth while, as it saves trouble and

Table 1a   Fuse elements for semi-enclosed
fuses (plain or tinned copper wire)

Current rating of fuse A	Nominal diameter of wire mm
3	0·15
5	0·20
10	0·35
15	0·50
20	0·60
25	0·75
30	0·85
45	1·25
60	1·53

mistakes in replacing fuses and costs nothing to reset when it has operated.

An important fuse that is too often overlooked is the small cartridge fuse in 13 A plugs and fused adaptors. This should always be checked when connecting appliance flexibles to ensure that only 3 A fuses are inserted for the smaller appliances.

Table 1(b) shows the correct size of cartridge fuse to fit into a 13 A plug for various domestic appliances in general use, but makers' instructions should always be referred to in this respect because some colour TV sets, vacuum cleaners, spin driers, heated washing machines, refrigerators and freezers of the larger kinds take high starting currents and may require the larger 13 A fuse. The normal loading of an appliance can generally be seen on the rating plate usually fixed to the back or base of the appliance.

Table 1b   Cartridge fuses suitable for 13 A plugs

3 A	*13 A*
Television	Electric fire
Radio	Convector
HiFi	Kettle
Blanket	Iron
Standard lamp	Toaster
Food mixer	Appliances between
Hair-drier	700 and 3000 watts
Refrigerator	
Sewing machine	
Vacuum cleaner	
Appliances up to	
700 watts	

Need for switches in live side

The incoming two-wire supply cable to a house consists of the live or line wire and the neutral wire. Switches make and break electrical contacts, thus closing or opening an electric circuit, and as there is a right and wrong method of connecting them Figs. 16(a) and (b) show alternative connections of a lamp and switch. In each case the switch will perform its function, but Fig. 16(a) shows the *correct method* and (b) is *wrong*. The invariable rule must be for all switches to be connected in the *live* wire, for which a *red* wire is always used. The reason is that the neutral (*black*) wire is connected to earth at the power station, while the live (*red*) wire is connected to the supply voltage, which may be as high as 250 V above earth potential. Thus in Fig. 16(a) the supply voltage *above* earth is disconnected from the lamp when the switch is open and it is safe. With the switch as shown in Fig. 16(b) the lamp is still connected to the full voltage, which is dangerous, despite the switch being open, if a person replaces a broken lamp thinking the lampholder to be 'dead'.

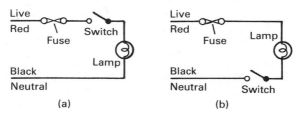

(a)   (b)

Fig. 16(a) Switch in live wire   **(b)** Switch in neutral wire (wrong)

Direct current and alternating current

Although d.c. for public supply is not now used in the UK it may be used in private installations on country estates and remote places to which the national grid has not penetrated; but it is good for the reader to understand the differences between a.c. and d.c. insofar as the electricity supply to a house is concerned.

A direct current is one that flows in one direction only, which is said to be from the positive ($+$) terminal or pole to the negative ($-$) terminal or pole. These markings will be seen on accumulators and some direct current instruments, which will only indicate when the positive wire is connected to the positive terminal and the negative wire to the negative terminal.

An alternating current does not flow constantly in one direction but reverses in a periodic manner. During one half of the cycle the direction is positive and during the next half-cycle it is negative. The number of complete alternations per second is called the *frequency*, denoted by 'f' or '~', and the unit is the *hertz* (Hz). There is no right and wrong way of connecting the two wires to a.c. instruments, as there is with d.c. voltmeters and ammeters.

Direct current can be represented by a horizontal straight line in the graph shown in Fig. 17, while alternating current is depicted by the wavy curve of Fig. 18. An alternating current of 1 A can do just as much work or light a lamp just as brilliantly as a direct current of 1 A. On a d.c. system the voltage has a steady value, while the alternating voltage varies in the manner indicated but has just as

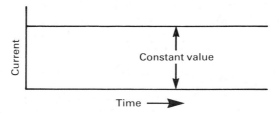

Fig. 17 Direct current

much effective pressure to send the current around the circuit.

Domestic apparatus that is marked 'Universal' for a certain voltage will operate equally well on either type of current; otherwise, it is specified as suitable for only one type. All appliances that depend on the heating effect, such as radiators, irons, kettles, etc., can be used on either a.c. or d.c. at their rated voltages. Apparatus depending on the magnetic effect will only work properly on the specified type of current, while in applications of the chemical effect d.c. is required. One can get an equally painful electric shock from either a.c. or d.c., though the effects vary.

Standard voltages and frequency

The term 'consumer's voltage' denotes the voltage at the incoming supply terminals, declared by the supplier. This corresponds to the term 'declared pressure' in the Electric Lighting Acts. Domestic

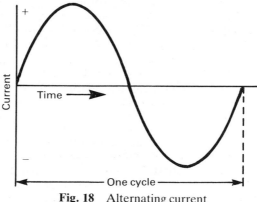

Fig. 18 Alternating current

consumers are supplied on low voltages and the standards for new systems and installations are:

Direct current systems (three-wire):
Consumer's voltage: 240 V and 480 V.

The former figure is that supplied for household use and is the voltage between the live conductor, which may be either positive or negative, and the neutral conductor. The latter figure is the voltage between the positive and negative outers (both 'live' conductors); it is used for large motors and large apparatus found in factories but not in the home. These direct current systems are virtually obsolete in developed countries but may exist elsewhere.

Alternating current systems (three-phase):
Consumer's voltage (declared): between neutral wire and each of the phase conductors, 240 V; between any two phase conductors, 415 V. Multiple earthed neutral.

The former value is used for domestic installations, while the latter figure is used for larger power purposes. Large buildings and institutions, houses and farms with large heating loads or machinery may have the incoming supply at the higher voltage, but it is divided into the lower voltage sections for the separate divisions and sub-circuits.

Frequency is now standardised in the UK at 50 cycles per second, or 50 Hz. This means that there are fifty of the waves shown in Fig. 18 in every second, or the time of one cycle is 0·02 second. The frequency does not affect apparatus depending on the heating effect, but small motors, transformers and any apparatus depending on the magnetic effect of a current will only operate satisfactorily at the correct frequency for which they are designed. The above figures apply to all the Electricity Board areas in the UK, but there may be systems existing in which the voltages vary from 100 to 250 and the frequencies between 25 and 100 c/s. Any supply undertaking will give details of its standard voltages and conditions of supply, as well as the different types of tariff available, together with any special local limitations on connected apparatus. These should be obtained for reference purposes for new installations.

Earthed neutral

The maximum single-phase pressure for domestic consumers is 240 V in the UK, and this is the voltage between any one line conductor and the neutral. The zero datum of the neutral is fixed by an earthed neutral point maintained by the supply authority. On alternating current systems the supply authority may provide a multiple-earthed neutral system, which means that the neutral conductor is permanently earthed at a number of points. Where one pole of a two-wire supply is connected to earth, this will affect the installation in a house since all fusing must be single-pole, i.e. single-pole m.c.b. or fuse assemblies are employed and single-pole fuseboards with neutral bar terminals for solid connection of the neutral circuit cables. There must be no break in the neutral conductor of the installation throughout the house. The earthing of the neutral is a safety measure that effectively prevents any part of the installation (exposed metalwork) becoming 'live' at a greater voltage than the single-phase voltage to earth in the event of a fault. Where neither pole of a two-wire supply is connected to earth, double-pole fuseboards must be used for both main and sub-circuits.

Protective multiple earthing (PME) is mainly adopted in the UK and is the system upon which installations described later in this book are based. But there are other systems which necessitate different principles in the method of earthing the installation and the form of protection to be adopted, and to which special Regulations apply. Therefore when designing an installation the particular earthing system of the supply authority must be provided for accordingly.

Medium and low voltages

For economic reasons, the supply authority distributes electricity at as high a voltage as possible, and street mains are usually at a pressure of 415 V between phase lines in a three-phase, four-wire a.c. system, or they may be at 480 V between 'outers' in a three-wire d.c. system. In both cases the neutral wire is generally at earth potential, and this limits the voltage between any line conductor and earth to 240 V. Thus the pressure of any live wire above earth is limited as a safety precaution. Up to 50 V a.c. or 120 V d.c. the voltage is now called extra-low; up to 1000 V a.c. or 1500 V d.c. supply voltage with up to 600 V a.c. or 900 V d.c. between any conductor and earth, is now called low voltage.

Effects of incorrect voltage

It is essential that apparatus of the correct voltage is installed, since if the supply voltage is too high damage may occur, while if it is too low the performance will be unsatisfactory. The voltage supplies the necessary impetus to send the current around the circuit to do the work required. Thus, if a 100 V metal-filament lamp is put in a lampholder connected to a 200 V supply, twice the proper pressure is applied and twice the current will flow through the lamp. Now, power (in watts) is proportional to the voltage times the current, so in this case the power applied to the lamp is $2V \times 2I = 2^2VI$, i.e. four times normal, and, of course, the lamp burns out. If, on the other hand, a 200 V lamp is put onto a 100 V supply, the current is only half that required, so the power is $(\frac{1}{2})^2$, i.e. a quarter of normal, and the lamp is very dull. Small differences of voltage have an effect on the illumination and the life of the lamp, both of which will vary considerably with the differences in the supply voltage. Thus, with a 1% increase of voltage on a metal-filament lamp, about 3½% more light is obtained, but the useful life is shortened by some 12%. With a 1% drop in voltage, there is about 3½% less light, but the useful life is lengthened by about 13%. Voltage drop in mains and circuits is, of course, the inevitable loss of pressure resulting from forcing current against the resistance of cables and wires. Supply voltage is limited by law to 6% variation above or below the declared voltage to allow for this, so that a consumer situated near a sub-station may have a slightly higher voltage than a remote consumer has, within these limits. Then there is the voltage drop in the consumer's installation itself which may be as much as 6 V at the end of the longest circuit in a large installation if the wiring has not been generous in design. This figure is limited by I.E.E. Regulations.

Incoming supply and methods of distribution

The provision of the supply is the responsibility of the supply authority which lays the service cable from the distributor running along the roadway. There may be a charge for opening up and connecting, and it is advisable to check that a large enough service cable is installed, though usually ample cable size is provided. The service cable is terminated at the supply authority's fuses, which are sealed so that they cannot be interfered with by unauthorised persons and are on the incoming side of the electricity meter. The supply then goes to the meter, and thence through the consumer's main switch to the distribution board. The connection and maintenance of the meter is the responsibility of the supply authority.

This arrangement is illustrated in Fig.19, which shows the service arrangement of a two-wire system. The service cables are connected to the distributors, which go back to the power station or sub-station. A three-wire system of d.c. distribution is shown in Fig. 20. The live outers are marked positive (+) and negative (−), and the

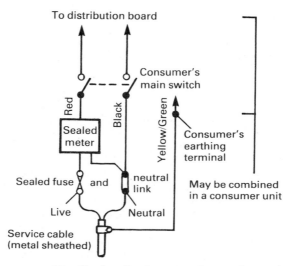

Fig. 19   Leading in a two-wire supply

earthed neutral ($\pm$) is shown in between the two outers. Two-wire connections to separate domestic consumers are shown at A and B; in each case, one conductor brought into the house is connected to a live conductor and the other one to the earthed neutral. Connection C shows the connections to a large consumer, the wiring being divided into two entirely separate parts and the load being approximately balanced on each side.

Fig. 21 indicates diagrammatically a three-phase a.c. four-wire system, which represents standard practice in the UK. The three-pointed star represents the transformer in the sub-station, the midpoint of which is earthed and connected to the neutral conductor. The three phase conductors form the live conductors, and are coloured red, yellow and blue. Inside the house RED is used for the live wire or phase conductor and BLACK for the neutral, the two different colourings being used throughout in A, B and C. With a large building the three-phase, four-wire supply may be brought in and divided up as shown at D. Separate consumers having two-wire connections are shown at A, B and C, the different houses being balanced along the road between the three phases. A large consumer might require to drive large motors; then the connections would be as indicated at D, so that the line voltage is 415 and the three voltages to neutral are each 240 V but in this case correct

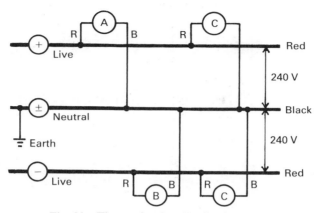

Fig. 20 Three-wire d.c. distribution

colours are used for the phase conductors, red, yellow and blue. In the diagrams the joints are shown by a black dot, and where two wires cross over without any connection the black dot is absent.

In farm installations the service line may be carried overhead on poles at high voltage, commonly 11,000 volts, terminating at a pole-mounted transformer with service fuses on the pole, and an overhead or underground cable, at supply voltage, to the consumer's switchboard where the meter would be situated.

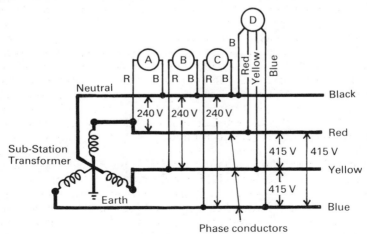

Fig. 21 Three-phase a.c. four-wire distribution

3

Transformers and Motors

Transformers are used on alternating current supplies to obtain changes of voltage and current with the same amount of power. A transformer is a static piece of apparatus consisting of a laminated iron core on which is wound either two separate windings, i.e. a double-wound transformer, or a single coil with a tapping point brought out, i.e. an autotransformer, as illustrated in Fig. 23.

A double-wound transformer is represented in Fig. 22, with an alternating current supply of 240 V applied to the left-hand coil, called the primary winding. An alternating magnetic field is set up in the iron core and it is this changing linkage of the magnetic field with the windings which induces an alternating voltage in the right-hand coil, called the secondary winding. With half as many secondary turns as on the primary winding, 120 V will be obtained across the secondary terminals. If this winding is connected to a load, the current will be approximately twice the value of the primary

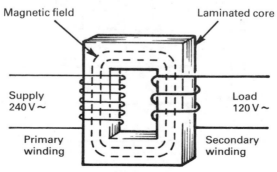

Fig. 22   Diagram of a double-wound transformer

current. Transformers have very small losses and the voltage is proportional to the number of turns in each winding, while the current is inversely proportional to the turns ratio.

Transformers with iron cores are used for low-frequency applications, i.e. on mains supplies and with low audio-frequencies in radio and television receivers, but at high radio-frequencies the iron core is omitted. This is because it impedes the rapid reversals of magnetism at high frequencies, but aids the magnetic circuit at low frequencies. Small double-wound transformers, with separate primary and secondary windings, for domestic bell and other extra-low-voltage circuits are mounted in either an iron or a bakelite case and should be properly earthed. They should comply with B.S. 3535 Class II and the secondary winding itself must not be earthed.

Transformers are only suitable for alternating currents, and if connected to a direct current supply of the same voltage they will take a higher current, get very hot and burn out, if the circuit fuse does not do so first.

Autotransformers

Autotransformers, which consist of a single winding on an iron core with a tapping at the required voltage, must not be used for extra-low-voltage bell and other circuits in the house, as the lower voltage winding is not entirely separate and insulated from the main supply. The lower voltage winding of an auto-transformer should be

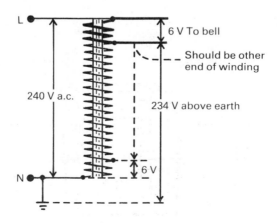

Fig. 23 Autotransformer incorrectly connected

adjacent to the neutral connection, and the common terminal of both windings should be connected to the neutral pole of the supply. Fig. 23 shows an autotransformer *incorrectly connected* and the dangerous condition it creates. The live conductor is shown connected to the common terminal for both sections, and it will be seen that, even though there is only 6 V, for example, across the bell circuit, this circuit is at least 234 V above earth. This voltage is dangerous both to the extra-low-voltage bell wiring and to any user of the equipment. N terminal should be the common terminal for both sides so that the extra-low-voltage connections are at the earthed neutral end of the winding.

Electric motors

For domestic applications electric motors are of small size and are usually fractional horsepower motors.

These small machines can be switched directly on to the supply but must be suitably wound for the type of supply and mains voltage. Direct-current motors will not operate on alternating current, and vice versa, in general.

There is one type, however, called the universal motor, that will run on either d.c. or a.c. of similar voltage but such motors do not exceed about 200 W.

Under the metric system, motor power output is measured in kilowatts or watts, and the equivalent rating of the many old ¼h.p. rated motors still in service is approximately 186 W.

Principle of the electric motor

An electric motor is used to convert electrical energy into mechanical energy and thus perform work at a certain rate. When a conductor is carrying an electric current, it exhibits magnetic properties. If another magnet system is adjacent to this conductor, there will be a mutual reaction and the wire will be forced to move in a certain direction. By a suitable arrangement of an iron electromagnet, which is generally stationary, and a rotating part, called the armature, which carries a suitable winding, mechanical power is obtained.

The outer stationary frame of an a.c. motor is called the stator, and the rotating part (termed the armature in a d.c. motor) is called the rotor. The main windings are on the stator, but the rotor may have a winding like a d.c. armature connected to a commutator or slip rings, or a number of bar conductors with a short-circuiting ring at each end, in which case it is called a squirrel-cage rotor.

Protection of small motors

The greater the power output of a motor, the more current it will take from the supply. If more work is demanded, the motor will take a greater current; if an overload is excessive, too much heat is generated and the speed of the motor may fall. To protect the motor and fuses, a thermal device or a small circuit-breaker may be employed, such as that shown in Fig. 36 on page 61, which opens the circuit when the current becomes excessive or an overload persists for too long. The fixed current setting gives protection that cannot be modified by unauthorised persons. The moulded enclosure affords protection from injury, and the 'free handle' mechanism prevents the circuit-breaker from being closed on a fault.

Types of d.c. motors

D.c. motors are mainly used for extra-low-voltage battery operations. Their type-names and characteristics are given as follows: the three main types of d.c. motors are (i) shunt, (ii) series and (iii) compound, depending on the arrangement of the field magnet windings. The shunt motor runs at practically constant speed, and its high-resistance field winding is connected in parallel with the armature winding.

The series motor has low-resistance field coils connected in series with the armature. The speed varies with the load, increasing as the latter falls, but it will exert a much greater turning moment, or torque, than the shunt motor at starting and low speeds.

The compound motor is a combination of these two types and is used when the load demands a higher starting torque than the shunt motor without the wide speed variation of a series motor. One application is driving the compressor of a refrigerator.

The connections of these motors are given in Fig. 24 while Fig. 25 shows characteristic curves of speed and torque against current.

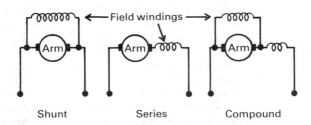

Fig. 24 Connections of d.c. motors

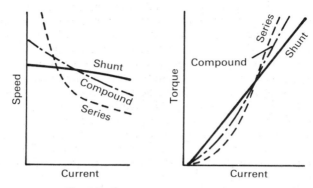

Fig. 25   D.c. motor characteristics

With d.c. motors the armature winding is connected at intervals to the commutator copper segments, which are insulated from one another and bear carbon brushes which form sliding contacts.

Wear of the brushes, dirty commutator and poor connections are the usual sources of trouble in small motors.

Universal series-type motors

These motors can be used on either d.c. or a.c. supplies of the same voltage. The entire magnetic system is laminated and the speed varies with load, as is usual with series motors. The performance is not as good on a.c. as on d.c., but it is quite suitable for vacuum cleaners, small fans, hair driers, floor polishers, mixing machines and other applications where the high or variable speed is acceptable. The armature is the same as that of a d.c. machine, but the brushes require more frequent renewal.

Types of a.c. motors

The a.c. series motor is similar to the universal motor and has already been described.

The basic principle of the static transformer, i.e. mutual induction between the windings, is the same principle that operates the induction motor, which can operate on three-, two- or single-phase supplies. Even though small three-phase induction motors have been developed, the single-phase induction motor is used for domestic applications, as only a single-phase supply is generally available. The single-phase motor is not inherently self-starting but will continue to run once it is started. In the split-phase type an auxiliary winding artificially makes the motor 'two-phase' during the starting period, while in the shaded-pole motor a copper shading ring around half the pole gives the necessary starting torque by causing time displacement of part of the magnetic circuit. This method can only be used for very light starting torque.

Split-phase motors are suitable for applications where heavy starting torques and overloads do not occur, e.g. electric washing machines fitted with a clutch. The artificial second-phase winding of a split-phase motor must be energised in a different period of time from the main single-phase winding, and to do this a capacitor is connected to the second- or starting-phase winding. This causes the current to be displaced from (to lead) the voltage, whereas the inductive windings of the motor cause the current to lag behind the voltage. Shaded-pole motors are used for small slow-speed fans. These motors have a 'shunt characteristic', i.e. the speed is nearly constant.

THE REPULSION-START, INDUCTION-RUN MOTOR This has one winding, connected to the supply on the stator, with an armature like a d.c. motor, but the brushes are connected together. It starts as a repulsion motor with a 'series characteristic', and when it has run up to some 75% of full speed the armature winding is short-circuited by means of a centrifugal device and contacts on the end of the rotor, and the brushes are lifted. It then runs as an induction motor.

THE CAPACITOR MOTOR The capacitor-start, induction-run motor has two windings on the stator, one of which is connected to a condenser or capacitor, which in effect makes it a two-phase motor for starting. The auxiliary winding and capacitor are cut out at a predetermined speed by a centrifugal switch and the motor runs as an induction motor. In the true capacitor motor, capacitors are switched into a different grouping and retained in circuit continuously after running up to speed, so that the motor runs under load as an artificial two-phase motor thereafter with enhanced performance. The rotor is of the robust squirrel-cage type; there is no electrical connection between the stator and rotor in either type. Although primarily developed for refrigerators and oil burners, these motors are extremely well suited to other duties requiring operation for long periods where maximum efficiency is essential. A capacitor-start motor is illustrated in Fig. 26. This motor has a resilient mounting. The robust rotor construction is evident from Fig. 27.

This is a most satisfactory type of single-phase, low-wattage motor, though it costs slightly more than the ordinary split-phase

Fig. 26   Capacitor motor with resilient mounting

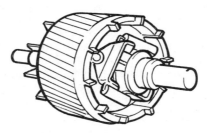

Fig. 27   Rotor of capacitor motor showing centrifugal switchgear on shaft

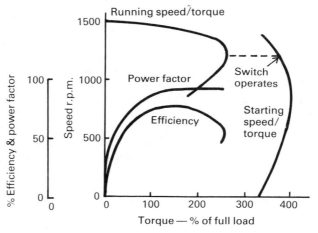

Fig. 28   Capacitor motor characteristics

motor. The relatively high efficiency and power factor is indicated in Fig. 28, together with the torque. Capacitor motors are inherently quiet, free from radio interference and economical in running cost.

SYNCHRONOUS MOTORS   Synchronous motors run at a constant speed, exactly in step with the supply frequency, whereas the motors previously described lose about 4–5% of the synchronous speed (which is called the 'slip') and are termed asynchronous motors. The domestic application is chiefly for electric clocks that operate on the 50 Hz (c/s) supply and are called synchronous clocks. One type requires the spindle to be given a twist to start. Another type makes use of the shaded-pole principle and is self-starting; the special shape of the rotor ensures that it runs at synchronous speed. The synchronous speed depends on the number of pairs of magnet poles on the motor and is independent of the voltage. Series motors and various a.c. commutator motors can run at speeds above and below synchronism. Table 2 below gives the synchronous speed for various numbers of poles on a 50 Hz (c/s) a.c. supply.

Table 2   Synchronous motor speeds on 50 Hz (c/s)

No of poles	2	4	6	8	10	12	16	20	24	30	40
Rev/min	3000	1500	1000	750	600	500	375	300	250	200	150

4

Primary Cell and Accumulator Batteries

A battery is a group of cells joined up to give greater power; the cells are generally connected in series, so that the individual voltage of each cell is added to the next one, as explained in Chapter 2. Cells of different sizes should not be connected together, since their internal resistances differ. Primary cells transform chemical energy into electrical energy. The Leclanché cell is generally used for bells and other small-current intermittent work; the chemical constituents are the same in both the wet and the 'dry' form. The wet Leclanché cell consists of a glass jar containing a solution of sal-ammoniac (ammonium chloride), in which stands a porous pot with a central carbon plate, which is surrounded by the depolariser. The terminal on the carbon is the positive, while a zinc rod in the electrolyte is the negative element. The electrolyte and the zinc rod are expendable and are easily renewed. Wet cells are best kept in a cool place to minimise evaporation; dry cells, on the other hand, are better in a slightly warm situation. The dry cell has the same chemical constituents as the wet cell, but the chemical paste, which takes the place of the sal-ammoniac solution, cannot be renewed; also, the outer zinc case becomes perforated, due to the chemical action, and cannot be replaced like the zinc rod of the wet-type cell. An inert cell is a form of dry cell that can be stored for long periods without deterioration. When required, a little water is introduced into a vent hole, thus moistening the chemical paste. The voltage of these cells is about 1·5 V; it falls if excessive current is taken from them, depending on their size, but they are quite satisfactory for intermittent use.

For continuous duty and heavier currents, secondary cells or accumulators are used. An accumulator has plates of certain chemical composition which is changed by the action of the charging current passing from the positive plate to the negative plate through the electrolyte. When charging is complete, the reverse action takes place with discharge, the accumulator supplies a current and the plates revert to their original condition. These processes can be repeated without renewing the constituents of the cell, which must be done with a primary battery.

There are several types of accumulator, but the most common types used for domestic purposes are:

(a) The lead-acid cell, with lead plates treated with certain oxides in a solution (electrolyte) of dilute sulphuric acid.
(b) The nickel-alkali cell, which has nickel-iron or nickel-cadmium plates with an electrolyte of concentrated potassium hydroxide.

New types of cell have been developed for electric vehicles, for the scientific equipment of spacecraft and for medical purposes. They include zinc/air, sodium/sulphur and lithium/chloride as electrode-active material combinations, but they are not all generally available yet, except the small alkaline dry cells which can now be readily obtained in the market place and have a much better performance in all respects than the old Leclanché type, particularly in output and length of life. The rapid development of small computers and other electronic control equipment has resulted in the development and production of the lithium anode primary battery with various cathode substances and solid or liquid electrolytes. Voltages of about 2 to over 3 volts per cell and long, reliable service life of over 5 years are obtained in some applications. Although some of the new types of cell are still only in the research laboratories, the lithium battery has been used for some time in circuitry for military, aeronautical and space exploration equipment; but there is now a growing need, in addition, for these small power, long-life primary cells in miniature electronic control gear, satellite communication systems, heart pacemakers, hearing aids, electronic wrist watches and portable measuring instruments, pocket computers, etc., which call for the characteristics of the lithium cell in particular.

Cells for small primary batteries are usually of the sealed type, having totally enclosed tops, but accumulators are provided with vent plugs or stoppers for access to the electrolyte. This reduces evaporation considerably and prevents spillage.

Lead-acid cell

The lead-acid cell is made up of pasted plates, the positive being red lead and the negative litharge mixed with sulphuric acid. An initial forming process changes these plates to chocolate-coloured lead peroxide and spongy lead of a grey metallic colour. The brightness of these colours is one of the indications that such a battery is well charged and in good condition. When the terminals are connected to an external load, the electrical potential existing between them forces a current to flow round the circuit, chemical changes occur, the plates go dull and the density of the electrolyte decreases, due to the formation of water. The density of the dilute sulphuric acid used varies with the type of cell but is generally around 1·21, and a measurement of the density with a hydrometer, illustrated in Fig. 29, is a good guide to the amount of charge left in the battery. Some accumulators have a floating indicator that shows the state of charge, while large storage batteries installed in power stations have hydrometers that float permanently in the acid.

The voltage of a freshly charged battery cell is about 2·2 V when it is not supplying current; but when discharging, it quickly falls to about 2 V, where it remains for some time, depending on the rate of discharge, and then gradually falls to such a low value that is useless and cannot support the current required. A lead-acid cell should

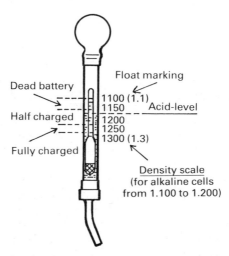

Fig. 29   Hydrometer for lead-acid accumulator

never be discharged below 1·8 V, otherwise insoluble lead sulphate forms on the plates. This is evident from white patches on the plates, which in time will ruin the battery.

Care of lead-acid accumulator
The following points, if observed, will prolong the life of a battery:

1. Never leave the battery in a discharged state. Charge the battery if the density falls below the minimum stated on the label (usually 1·150).
2. A voltmeter reading across the terminals, an open circuit, is deceptive; it may be 2 V, but when discharging it may fall below 1·8 V.
3. Keep to the recommended charge and discharge rates specified by the maker.
4. The level of the electrolyte must be kept above the top of the plates by adding distilled water to make up for evaporation. Concentrated sulphuric acid should not be added. In mixing the electrolyte, the acid should always be added to the water; the other way round is dangerous.
5. All connections should be kept clean and well smeared with vaseline.
6. Do not short-circuit the battery with a piece of wire; this is dangerous and a heavy short-circuit current will loosen the paste and may buckle the plates.

Nickel-alkaline cell

The positive plate in each case is nickelic hydroxide held in a grid of nickel, while the negative plate is a special mixture of iron oxide or cadmium carried in nickelled steel plates. The electrolyte is potassium hydroxide (caustic potash), which is a corrosive liquid with a specific gravity about 1·18. The electrolyte does not change in density, so only very occasional make-up with a little distilled water is necessary.

The average voltage of an alkaline cell is 1·2 V, so more cells are required to make up a given battery voltage than lead cells with voltage of 2 V. The advantages of this cell are its light weight, ability to withstand mechanical abuse and freedom from 'sulphating', and that it can be overcharged, overdischarged and left uncharged for long periods without damage. The disadvantages are higher initial cost and the need for more cells to make up a given voltage.

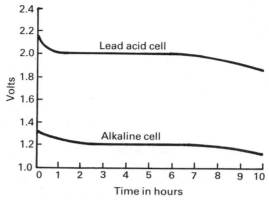

Fig. 30   Discharge characteristics of lead-acid and alkaline cells

Capacity of a battery

This is usually given on the label in ampère-hours at the ten-hour rate, e.g. a 40 Ah battery means that it will supply 4 ampères for ten hours, if in good condition. With higher discharge rates, the capacity is reduced. With a lead cell, if the current in this example is increased to 6 A, it will only last for six hours before the p.d. falls to 1·8 V, and with currents below 4 A the ampère hours obtained are somewhat above the rated capacity. The number of ampère-hours on discharge compared with that required to recharge the cell (ampère-hour efficiency) is about 90% for a lead cell and 75–80% for an alkaline cell, though the latter has less difference in capacity with different discharge currents. Comparative characteristic curves are given in Fig. 30.

Methods of charging accumulators

The notes on simple circuits and connection of battery cells are given in Chapter 2, so they are not repeated here; however, they may usefully be reviewed by the reader, as recognition of polarity is important with accumulators. The positive terminal of the charging circuit must be connected to the positive terminal of the battery and, likewise, negative to negative. Alternating current cannot be used directly for charging batteries, because an accumulator requires a continuous direct current to charge it whereas alternating current reverses its direction in each alternation; so it has to be 'rectified' –

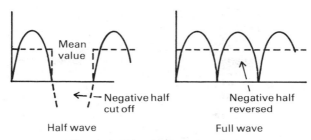

Mean value

←—Negative half cut off

Negative half reversed

Half wave

Full wave

Fig. 31 Illustrating rectification

i.e. only the positive half of the wave is fed to the battery, or by more complete rectification the negative half is brought over to the positive side, as indicated in Fig. 31. This unidirectional current is quite satisfactory as the charging current and, although not strictly continuous, has an effective value equal to the mean value of the current half-waves.

Accumulators can be charged from d.c. supplies by inserting a resistance in series to 'drop' the difference of voltage, with the required charging current. But this method is wasteful, as the following example will show.

EXAMPLE 9.   A 12 V car battery is charged at 15 V 5 A from a 110 V d.c. supply. What resistance must be connected in series? What will be the loss in the resistance and the cost of charging for 24 hours, assuming the mean charging current is 4 A and electricity costs 6p per unit?

Volts drop in resistance $= 110 - 15 = 95$ V.
Value of resistance $= 95 \div 5 = 19 \, \Omega$, to carry 5 A.
  Loss in resistance (I^2R) $= 5^2 \times 19 = 475$ W
  (compared with 15×5 $= 75$ W input to battery).

Cost per 24 hours $= \dfrac{110 \times 4 \times 24}{1000} \times 6p = $ say 64p.

A better method is to use a motor generator set, consisting of either an a.c. or a d.c. motor driving a low-voltage generator with a suitable output for the batteries to be charged. But this method is only suitable for large or groups of batteries where skilled attention is available.

Houses in remote places without mains supplies may have private generating plants with petrol-engine-driven d.c. generators and large-capacity accumulator batteries which can supply lighting etc. in the house for long periods between charging periods. Such installations must have proper switchboards with instruments and automatic charging equipment.

Large establishments where the public is admitted and other places where maintenance of supply is essential if the main supply fails are usually provided with large-capacity batteries for emergency stand-by supplies. These batteries usually supply an obligatory minimum degree of lighting for a few hours and are generally 'trickle' charged by charging equipment from the mains to keep them in fully charged condition ready for an emergency. These larger battery installations require skilled maintenance and attention to keep them in good working order; they are not treated in further detail as they are considered to be outside the scope of domestic installations.

When the lead-acid cell is being charged, it 'gases' and naked lights or sparking connections or contacts must not be allowed near the battery, as the hydrogen evolved forms an explosive mixture with air which is a serious risk with large batteries.

The car battery in service is charged by a special type of generator driven by the engine, and the circuit includes a 'cut-out', which completes the charging circuit when the voltage of the dynamo is high enough to provide a charging current and switches off when low engine speed causes the dynamo voltage to fall too low, so preventing the battery from discharging through the dynamo. In the modern car an a.c. generator is often fitted instead of a dynamo, with an electronic rectifier to provide direct current for battery charging and having the advantages of the a.c. supply for other purposes.

For the small accumulator and car battery, static types of charger are most suitable. These are usually of the valve-rectifier type for the larger heavy-current chargers used in garages and works where battery-driven trucks or trolleys are employed, and of the dry-plate rectifier type for small light-current and 'trickle' chargers commonly used by car owners for keeping their car batteries charged when out of use for several days. The equipment consists of a transformer to reduce the a.c. mains voltage to the charging voltage, and the secondary circuit includes a rectifier pack in the circuit to the output terminals. A fuse is fitted in the primary circuit and a thermal cut-out is usually included in the output circuit to protect the rectifier and windings in the event of short-circuit. The small car battery chargers have a screw-plug adjustment for the mains voltage side for a range of a.c. voltages from 200 to 250. On the output side, similar adjustment is provided for 2 V, 6 V and 12 V batteries. The charger is usually supplied with a length of cable and spring clips to clamp on to the battery terminals.

The cost of charging a car battery in this way is very different from that given in Example 9 because the losses in the charger transformer and rectifier are comparatively small. The electricity used to completely charge a discharged car battery, similar to the battery in the example, from 240 V mains in this way would only amount to less than two units.

Emergency lighting units

Developments in electronics and solid-state circuitry have enabled a new application of small secondary cells to be made in the independent, self-contained emergency lighting fittings that are now available. Although intended for the safety of the public in stores, schools, hospitals, etc., there is no reason why they should not be used in a private house if thought desirable.

A small diffusing bowl encloses two small lamps, a sealed three-cell, nickel-cadmium, alkaline battery, a trickle charger and mains failure relay in a remarkably small space. With a power consumption of only 5 W from the mains, it automatically provides a small amount of light, enough to find one's way about, for a period of one to three hours if the mains supply breaks down, and practically no maintenance attention is required. Larger units are also available and some contain miniature fluorescent lamps.

5

Switchgear, Distribution and Wiring

The main fuses protect the whole installation and may be housed in a separate case or included with the main switch.

Modern practice is to use either an iron or a moulded insulation case for the main switch and fuse(s). The main switch is of the double-pole type, so that the installation is completely disconnected from the supply when the switch is in the 'off' position. With large a.c. installations, a triple-pole switch is used to connect the three-phase supply to the distribution board or boards to which a single-phase supply is given through separate pairs of cables, made up by one connection from each phase and the common neutral link.

The main fuses are contained in porcelain or moulded insulation carriers and the fuse wire is completely protected from accidental contact or cartridge fuses may be fitted. The cover of combined switch fuses is provided with a mechanical interlock, which prevents the cover being opened unless the operating handle is in the 'off' position. It is then safe to withdraw the fuses for replacement or to isolate the installation from the supply without the possibility of making or breaking the circuit on the fuse-carrier contacts, or of coming into personal contact with live parts. The iron case must be earthed, and a terminal is provided for this purpose. The earthing wire is an important requirement for safety throughout the installation and details of the sizes of this protective conductor are given under the heading *Earthing and Equipotential Bonding* at the end of Chapter 6. Standard sizes for main switches suitable for most house installations range between 30 A and 100 A.

Whilst modern designs of domestic switchgear include a white or brown insulating case, many existing installations will have the

Fig. 32 250 V 30 A Single-pole and neutral all-insulated switch-fuse showing base with cover removed

(*MEM Ltd.*)

much older metal-clad units, but the interior arrangement of components and electrical connections are very similar. An all-insulated single-pole and neutral main switch and fuse is shown in Fig. 32.

Splitter switches can be employed for small installations. These combine the main switch and fuses for the separate circuits. Such an arrangement is inexpensive, but it is only adopted for small installations where there are very few sub-circuits. Splitter switches usually contain a main switch and two or three 15 A single- or double-pole fuses in one case for two-wire installations. There is no separate consumer's main fuse. A splitter switch with two single-pole and neutral ways is shown diagrammatically in Fig. 33. Three-way, metal- and moulded-cased splitter switches are shown in Fig. 34.

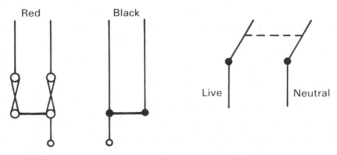

Fig. 33 Two-way switch splitter diagram

(a)

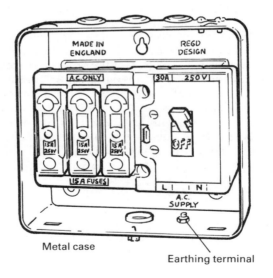

(b)

Metal case

Earthing terminal

Fig. 34   Three-way, 250 V, 45 A splitter switches
(**a**) moulded case (*Delta Ltd.*)
(**b**) metal case, cover removed (*MEM Ltd.*)

A development of the splitter unit, known as the consumer control unit, is commonly used in modern house installations. This unit combines a 60 A or 100 A double-pole main switch with single-pole circuit fuses or miniature circuit-breakers (m.c.b.s) and a neutral terminal bar in one case. The consumer's main fuse is not necessary with such an arrangement, and the circuit fuses or m.c.b.s. are designed to suit the usual house circuits, i.e. up to

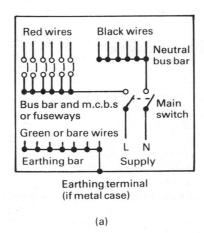

(a)

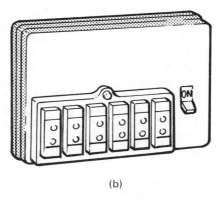

(b)

Fig. 35   A typical six-way consumer control unit with miniature circuit-breakers

(Wylex)

twelve or more ways of varying capacities between 5 A and 45 A for lighting, socket outlet, cooker and similar circuits. Fig. 35 shows a typical consumer's unit with m.c.b.-ways. In modern house installations consumer units have almost completely superseded the main switch and fuse gear previously described and which are usually only found in old premises nowadays.

In a switch-fuse the switch and the fuses are constructed separately but in a fuse-switch the fuses are mounted on, and form part of, the switch movement.

Miniature circuit-breakers

Small-capacity, automatic circuit-breakers for the protection of final circuits are now commonly used. They are grouped in distribution panels and consumer units in the same way as fuses, with similar or various ratings to suit the capacity required in the final circuits. The size is the same as a fuseway up to 45 A capacity, and some manufacturers make basic distribution units into which rewireable or cartridge fuse-carriers or miniature circuit-breakers can be plugged as required without altering the design of the distribution unit. The advantages are elimination of fuse rewiring or replacement – to say nothing of replacement by the wrong size of wire or cartridge – visual indication of the opened circuit and the 'free handle' action, which prevents closing the circuit if the fault persists. The circuit-breakers are sealed for safety and thus are immune from tampering. These miniature circuit-breakers are tested to break an a.c. short-circuit current of several thousand ampères, and give good protection and discrimination by thermal operation on overload and a magnetic trip under heavy fault conditions. Such a breaker is shown in section in Fig. 36 (page 61).

High speed miniature circuit-breakers can operate before the maximum value of short-circuit current has been reached thus reducing the potential damage at the fault location.

Concern for electrical safety has developed rapidly in recent years and has led to a trend to utilise residual current circuit-breakers instead of the ordinary miniature circuit-breakers in distribution boards for selected circuits where the risks are abnormal, or instead of the main switch in consumer units. They can now be readily supplied by several manufacturers; the space required is about the same.

The increase in safety of the installation in which they are used is incalculable, because a fault will cause the circuit or installation to be isolated so rapidly that a shock voltage cannot be maintained long enough to become lethal. This aspect is dealt with more fully under *Operation of residual current devices* in Chapter 6, p. 104.

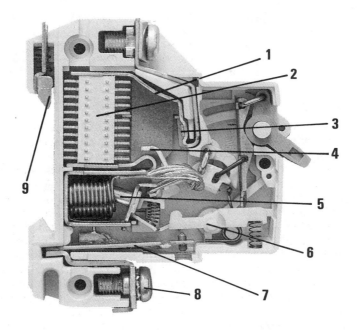

Fig. 36 Mechanism of miniature circuit-breaker

(*Crabtree*)

1. Arc runner.
2. Arc chamber.
3. Fixed contact.
4. Moving contact.
5. Hammer Action solenoid.
6. Trip bar.
7. Thermo-metal.
8. Wiring terminal.
9. Fixing by clip-on rail mounting.

Main circuits and final circuits

The main switch and fuse control the total current demanded by the premises. The total current can be compared to a river that gets its

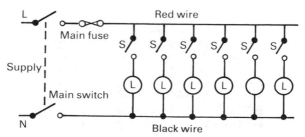

Fig. 37   Simple final circuit of six lamps only

total flow from a number of tributaries, but the flow is reversed. The final circuits are the tributaries, and each one must be protected by its own fuse, so that if one section develops a fault it is cut out and isolated without the whole installation having to be disconnected.

In Fig. 37 six lamps are shown connected to one fuseway, in which L indicates a lamp and S a switch. It should be noted that the switches are connected to the RED WIRE on the live or line side, while the lamps are connected to the BLACK WIRE on the earthed, neutral side.

With larger installations the various circuits are supplied from a consumer control unit (Fig. 35) or a switch fuse and separate distribution board, the cables being suitably routed for convenience and economy. In large rooms two separate circuits may be used, so that if one circuit develops a fault there is enough lighting left for safety. Socket outlets rated at 13 A are given separate radial group circuits or are arranged on a ring circuit.

If electrical energy is supplied through two meters, say one for lighting and the other for power, then entirely separate main circuits must be run with no possibility of interconnecting the different types of load.

The main distribution board is supplied, from the main fuse to the bus-bars, with cables large enough to carry the total current (and with some margin for future increase of load) if the distribution board is remote from the service position and main switch fuse. But in the typical house, where there is no need to separate this equipment, a consumer unit is usually provided; the supply authority's service fuse is permitted to be used for the protection of the whole installation and the cables are taken directly into the consumer unit. Such an arrangement is shown in Fig. 38 for a consumer unit with eight ways. There are two lighting circuits to lamps with their associated switches; one radial and one ring circuits to socket

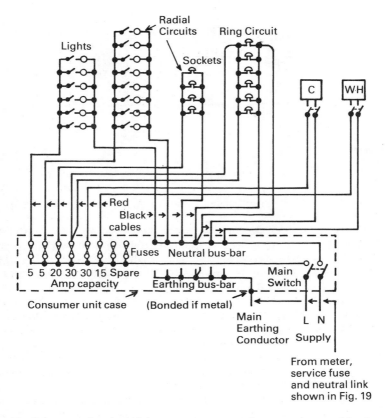

Fig. 38 Diagram of typical eight-way consumer unit connections and final circuits

outlets; one circuit for a cooker and one for a water heater; and two spare ways for possible future additions. This is the most convenient arrangement for medium-sized houses when all current passes through one meter, with an 'all-in' tariff. One distribution board is sufficient, but lighting and power circuits are separated because of the different current capacities.

With separate scales of charges, two separate distribution boards would be required and the installation would be more expensive. These arrangements are shown by a single-line diagram in Fig. 39. Water heaters and cookers are also run on separate circuits, some-

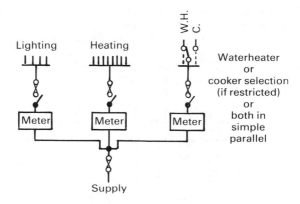

Fig. 39   Single-line diagram to illustrate separate metering

times with separate meters, main switches and fuses. Subdistribution boards are necessary in large establishments with long cable runs, where the distance from the main distribution board would be too far for some final circuits due to excessive voltage drop. A sub-board is fed from a pair of cables brought back to the main distribution board. If this sub-board is connected directly to the main switchboard, its own switch fuse gives protection to the cables; but if it is connected to a main distribution board, the correct practice is to connect it to a fuseway, as indicated in Fig. 40.

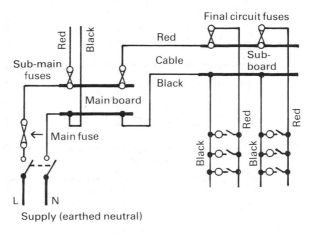

Fig. 40   Sub-board fed from fuseway on main board

For buildings with several floors of large extent, a suitable arrangement is shown in Fig. 41, in which each pair of rising cables is connected to a fuseway on the main distribution board.

In tall tower blocks with many floors it is common practice to install rising mains, of suitable capacity and of uniform section

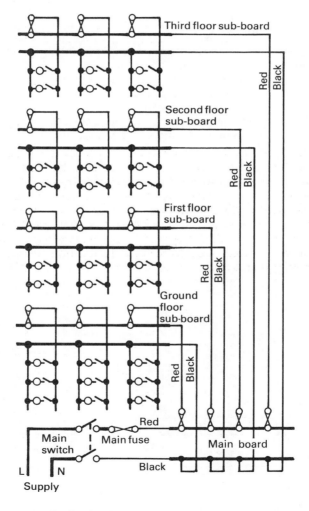

Fig. 41 Main distribution board with sub-boards connected by fuses

copper bars in a protective casing, which are run vertically to the top floor of the building with service cables tapped off at each floor level and run to individual flats. In such cases no fuses are fitted except at the services in the flats and the rising mains installation is usually under the control of the supply authority. This is, in effect, like the street mains in a road with houses.

Installation systems

The various wiring systems employed for installations are generally designed for economy, length of life or mechanical strength, depending on the class of work required and the nature of the building. With any system, the avoidance of danger to life and the safety of the structure should take precedence over other considerations.

In Table 3 below various wiring methods are listed in order of merit, but the comparisons must not be taken too literally.

1 Conduit systems

These are the most satisfactory systems, because rewiring can easily be done, if the system is properly designed, without disturbing the building's structure and finishes. If rewiring is to be provided for,

Table 3   Wiring methods compared approximately

System	Relative %			Labour required	Extensions and renewals	Protection against		
	Cost	Life	Time to install			Fire	Damp	Mechanical damage
1 Conduit (metallic) (plastic)	100 75	100 100	100 75	Skilled Skilled	Difficult Fair	Good Good	Good Good	Very good Good
2 Mineral insulated metal-sheathed	150	150	90	Skilled	Fairly easy	Very good	Good	Good
3 P.v.c.-sheathed	60	90	50	Semi-skilled	Easy	Poor	Fair	Fair
4 Wire Armoured P.V.C. insulated	100	100	75	Skilled	Difficult	Good	Good	Very good

elbows and tees should not be used, since they prevent the easy drawing in of new cables when rewiring is necessary. Site formed easy bends with accessible junction boxes or inspection boxes should be used. Steel tubes with screwed joints can be concealed in the walls during building; for surface work they provide the best mechanical protection for cables; and by suitable precautions it is possible to make the wiring practically waterproof. Light and heavy-gauge conduit is available, which may be solid-drawn seamless or seam-welded with an almost imperceptible joint. The heavy-gauge welded tube is used for high standard screwed work; light-gauge welded conduit is used for cheaper work. Steel conduit is generally finished in black enamel, inside and out, but for very damp sites galvanised or zinc-impregnated finish is used. Gas barrel is fitted for some outside work; it is thicker than conduit and adaptors are required to connect the two. Gas barrel sizes are given in inside diameters, while conduit is specified by its outside diameter. Special fittings are necessary with conduit, consisting of elbows, tees, bends, junction boxes and inspection boxes illustrated in Figs. 42 and 43. The insulated wires are drawn into the conduit after it has been installed, but the tube must not be packed tight with the wires to avoid damaging them and there should always be room to draw in additional conductors. It is possible to accommodate a greater number of wires in a short straight run than when there are a number of bends and junctions.

Comprehensive tables of factors are provided in the Regulations for easy calculation which take account of bends and sets in the run

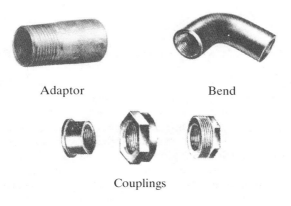

Adaptor   Bend

Couplings

Fig. 42   Some screwed conduit fittings.

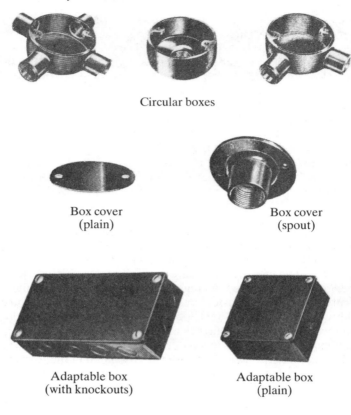

Circular boxes

Box cover
(plain)

Box cover
(spout)

Adaptable box
(with knockouts)

Adaptable box
(plain)

Fig. 43   Some conduit boxes.
Iron, steel or plastic boxes are similar in shape and usage.

of conduit and length of run where cables have to be pulled in. A few representative samples for sizes commonly used in house wiring are given in Tables 4 and 5, but the factors for all conditions and numerous lengths of run are too extensive to include. Similar tables and calculations apply for the capacities of standard trunking, but this is not usually found in domestic installations.

The wiring capacities of conduits are found by using factors from Tables 4 and 5. The factor for each cable is added to obtain a total for the capacity of the conduit and the conduit size is found by selecting the conduit with the same or next higher factor number.

Table 4 Cable factors for single core p.v.c. insulated cables in steel or plastic conduit (per wire)

Section (mm²)	Short straight runs up to 3 m		Straight runs exceeding 3 m or any length with bends and sets
	Stranded conductors	Solid conductors	
1·0	—	22	16
1·5	31	27	22
2·5	43	39	30
4·0	58	—	43
6·0	88	—	58
10·0	146	—	105

Thus three 4 mm² (7/0·85) cables in a run of 4 m require 16 mm conduit with not more than two bends, i.e. $3 \times 43 = 129$ which is within the factor of 130 for 16 mm conduit.

The effect of bunching a number of wires together is to reduce the amount of current each wire can carry, so that when the choice of cable is made to suit the load current and permitted voltage drop in the circuit (see Table 7, page 85) a further reducing factor has to be allowed for the number of cables to be bunched together in the conduit or trunking as this reduces the natural cooling of the cables. This factor may vary from 0·8 for four similar cables to 0·59 for

Table 5 Factors for capacity of conduits

Conduit dia. (mm)	No. of bends	Short straight runs (0–3 m)	Length of run (m)						
			1	2	3	4	6	8	10
16	0	290	—	—	—	177	167	158	150
	1	—	188	177	167	158	143	130	120
	2	—	177	158	143	130	111	97	86
	3	—	158	130	111	97	—	—	—
20	0	460	—	—	—	286	270	256	244
	1	—	303	286	270	256	233	213	196
	2	—	286	256	233	213	182	159	141
	3	—	256	213	182	159	—	—	—
25	0	800	—	—	—	514	487	463	442
	1	—	543	514	487	463	422	388	358
	2	—	514	463	422	388	333	292	260
	3	—	463	388	333	292	—	—	—

ten and 0·36 for forty loaded cables according to the degree of bunching.

Plastic conduits are increasingly being used for installations in housing and commercial premises, though not in factories. Table 3 shows them to compare favourably with other good-quality systems. Plastic conduit is much more flexible than steel conduit; it is more easily bent to shape; and although the heavy-gauge tubing can be screwed, if required, it can simply be cemented into the spouts of the associated accessory fittings and boxes. Some manufacturers provide special grip outlet fittings and boxes for plastic conduit. If a luminaire (the modern international term for a light fitting) is suspended from or is in contact with a plastic or non-metallic outlet box the weight of the fitting must not exceed 3 kg nor the box temperature 60°C. Increasing use is now being made of hollow skirtings and door architraves which are designed to provide accommodation for wiring and to enable outlets to be fitted at any point in a room. This is a great convenience for later alterations, additions and extensions to the installation. Copper and aluminium conduits are used where the advantages of one of these metals over steel is of importance, but aluminium has the advantage over copper of cost and labour-saving features of lightness and ductility.

Most of what has been said about steel conduits also applies to plastic, copper and aluminium conduits and their fittings. In the standard code of colours for services pipework electric conduits are coloured orange for identification.

CONDENSATION IN CONDUIT   In damp atmospheres where changes of temperature occur there is the possibility of moisture condensing inside the conduit. When the conduit is warm the air inside it expands and, on cooling, air and moisture are drawn into the conduit. This 'breathing' action can be minimised by making watertight joints at the cable boxes and switches with compound. The necessity for good continuity in metal conduit systems presupposes that all joints are clean and metal-to-metal. It has been found that red-lead and tow joints give almost as good electrical continuity as dry joints and have the advantage that they are watertight. Metallic paint is also commonly used to make sound joints. Where the conduit is buried and cement is floated on or much wet plaster is used, these methods are important for preventing the ingress of water into the conduits. The moisture tends to rust the inside of the conduit and is likely to be more troublesome on light-gauge than on heavy-gauge conduit, especially if the enamelling is poor. With long vertical runs, an open T-piece is sometimes fitted at the bottom to allow drainage and some ventilation, while horizontal runs are given a slight slope to the lowest point, where a through box or T is fitted, or a drain hole is provided. Rusty patches and moisture also tend to affect the rubber insulation of cables. For this reason, conduit systems erected during building should be given some time to dry or be swabbed out and lubricated before cables are drawn in.

EARTHING OF CONDUITS   Metal conduit systems must be continuous electrically and mechanically, and must be effectively connected to earth because they are normally used as the protective conductor of the installation.

Where earthing clips are employed for steel conduit systems; they must be put on bare metal, so the enamel has to be scraped off the conduit. They consist of tinned brass or copper with a terminal or lug, to which the protective conductor is screwed or soldered. An earthing terminal is provided on metal-cased apparatus. Continuity and earthing are safety precautions, so that in the event of the conduit or other exposed metalwork becoming 'live' the fault current can leak to earth without a dangerous 'touch' voltage arising to cause shock. With a 'dead earth' or short-circuit the fuse will blow, isolating the faulty circuit, while small faults may only show up on an insulation test. With poor joints the contact resistance will be high, and may constitute a fire and shock risk if the fault current is not sufficient to operate the circuit protective device; it is therefore necessary that the earthing circuit (see Fig. 62, p. 102) impedance must be sufficiently low for the operation of protective devices (fuses or circuit-breakers) within limited periods and before the prospective shock voltage can occur between exposed metalwork and other neighbouring parts and produce conditions for possible electric shock. Therefore where conduits are used for the protective conductor it is required that their resistance values must comply with the limits set for any other protective conductor, further details of which are given under the heading *Earth fault loop impedance* in chapter 6 (page 100).

2 Sheathed wiring

Originally, lead or rubber covered wiring having the conductors insulated with vulcanised rubber, the cores being red and black, with a sheath of extruded lead alloy or vulcanised rubber, was the most popular method of installation, but p.v.c. (polyvinyl chloride compound) sheathed cables are now the most popular and cheapest form of wiring for domestic installations. The conductors of p.v.c.-sheathed cables are insulated with red and black p.v.c. to distin-

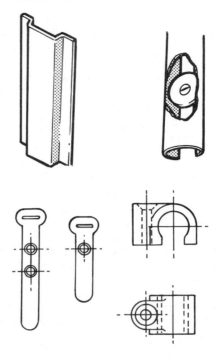

Fig. 44   Plastic channelling, plain buckle clips and nylon clip for sheathed
cables

guish phase and neutral wires; and two or three cores, placed side by
side, with or without a bare protective earthing conductor, and
sheathed overall with p.v.c., are commonly used for house wiring. It
is important to use an ample number of fixing clips to avoid sagging
of the cable and to preserve a neat appearance. Although p.v.c. is
very tough, it is vulnerable to mechanical damage. As protection,
oval conduit or a channel of wood, metal or plastic may be used for
running down wall surfaces, but p.v.c. insulated cables with a p.v.c.
outer sheath may be installed directly within walls, cavities or under
plaster. Plastic channelling and cable fixing clips are illustrated in
Fig. 44. The buckle clips are fixed by nails and bent over the cable,
the end being threaded through the eye. P.v.c. cable should not be
placed in contact with expanded polystyrene insulation or other
thermoplastics as p.v.c. is softened by the chemical plasticisers.
Modern cable clips are made of nylon or hard plastic and designed

to hold the cable when it is pressed into place. Fixing clips and saddles for cables and conduit have generally been fixed to building surfaces by means of wall plugs in masonry and nails or screws, but the latest ideas which may, in time, supersede these old and tried methods use powerful quick drying adhesives. Such adhesives are already commonly used for joining plastic conduits and accessories with plain spout entries. A junction box for p.v.c.-sheathed cables is shown in Fig. 45. Harness systems consist of a central junction box with suitable lengths of cable to reach all the outlet points in a house and are factory-made ready to install, thus saving much work on site.

In all forms of sheathed wiring it is important to:

(a) Connect all earthing conductors together, and to an earthing terminal in the junction box, switch box or other outlet access-ory where each cable is terminated.

(b) Take the sheathed cable end right into the outlet enclosure or junction box so that the cable cores, where unsheathed, are contained within and not exposed outside the outlet enclosure or junction box.

(c) Protect fixed cables where run on wall surfaces in vulnerable positions (up to, say, 2 m above the floor) from impact damage by means of wood, metal or plastic channelling.

There are various patented systems using special fittings which make for ease in erection and safety in operation. Skilled labour is

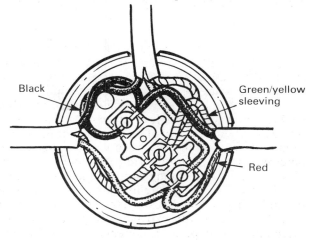

Fig. 45   Junction box for p.v.c-sheathed cables

required to make a neat job. Cable should be carefully run off the drum, as kinks are very difficult to get out once they occur and are liable to fracture the conductors. Care must be taken to avoid sharp bends and undue pressure on plastic sheathing. It is advisable to keep these types of cable out of direct sunlight to avoid softening.

In multi-core cables the earthing conductor is usually bare within the sheathing, but where it is exposed at terminations it should be insulated with green-yellow sleeving. Although insulated lamp-holders and insulated switches do not need earthing, it is a rule to provide an earthing terminal in all outlet boxes so that any exposed metalwork of future fittings and switches can be earthed.

P.v.c. cables are tough and flexible at normal operating temperatures, highly resistant to burning, chemically inert and unaffected by sunlight. As p.v.c. is a thermoplastic it should not be used under heat and pressure. For conduit work p.v.c. insulation moulded directly on to the wire does not need further protective sheathing.

PLASTIC TRUNKING   Plastic trunking can form a neater surface enclosure for cables than large conduits, especially where employed as room skirtings, architraves or cornices. Hollow trunkings of this kind make wiring easily accessible and available on any wall of a

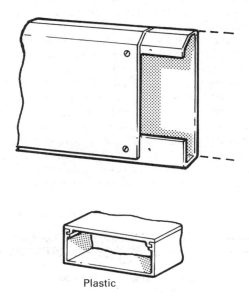

Plastic

Fig. 46   Plastic trunking, skirting and rectangular types

room for additional socket outlets and extensions. Skirting and rectangular trunking is illustrated in Fig. 46.

Grey or white rectangular-section plastic cable trunking of various sizes with clip-in covers is made to contain p.v.c.-insulated single wires; it is often used in conjunction with a plastic conduit system. The walls should be plugged and the casing secured by screws. The capping is usually slipped into position by springing the trunking open to allow the tongued and grooved edges to engage.

3 Mineral-insulated cables

This type of wiring consists of single- or multi-core, solid-strand conductors surrounded by compressed magnesium oxide insulation with an external metal sheath and formed into a solid-drawn cable. The conductors and sheath were originally of copper, but because of the high cost of copper, aluminium is now used also. An outer p.v.c. sheath is added where the metal sheath would be liable to corrosion. This cable is very heat-resistant, even fire-resistant, to a degree beyond all other cables, and can suffer appreciable deformation without electrical failure. The ends are liable to absorb moisture, so they must be sealed with special accessories to obtain a watertight installation. Fig. 47 shows a typical cable termination with a screwed compression gland. Current ratings for mineral-insulated cables, not p.v.c.-sheathed, are much higher than for p.v.c.-insulated cables, which means that they can operate at higher temperatures, but voltage drop in long runs is consequently a more important consideration. A mineral-insulated cable was allowed to reach a temperature not exceeding 250°C (482°F) provided that it was out of reach and not likely to be injurious to its surroundings (150°C for terminations), when it was first introduced, but current-carrying tables for this type of cable do not now envisage situations where the sheath temperature exceeds 70°C for light duty cables exposed to touch or p.v.c. sheathed, or 105°C with bare sheath and not exposed to touch nor in contact with combustible materials. No correction factors are applied for bunching of mineral insulated cables. These

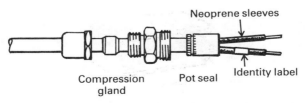

Fig. 47   Mineral-insulated cable termination

cables are not suitable for discharge lighting circuits unless precautions are taken against voltage surges.

4 Armoured cables

Where more protection is required than can be provided by a p.v.c., rubber or paper insulated and p.v.c. sheathed cable or where the use of conduit or trunking is not appropriate, especially for heavy current ratings and cables for vulnerable routes or to be buried in the ground, multicore p.v.c. insulated, metal armoured and p.v.c. sheathed cables are used. These cables may have two or more cores of all sizes from 1.5 mm^2 to 300 mm^2 and can be armoured with metal braids, wires or tapes of materials suitable for the protection required but for most purposes are provided with a closely laid layer of galvanised steel wires under the outer sheath. The core colours are red/black; red/yellow/blue; red/yellow/blue/black for two, three and four cores respectively and may be numbered or colour coded for extra cores but a green/yellow earthing core is not included. The armoured cables are terminated with cable glands designed to grip the armouring and to ensure that earthing continuity is maintained, thus under normal circumstances a separate earthing core is not required. An armoured cable is shown in Fig. 48.

This cable, where run underground, must be free of damage from stones and sharp objects and be buried at a depth below normal cultivation levels with protective tiles or concrete slabs above it, or installed above ground supported by cable ties, clips or cleats. It is completely watertight, robust and for large current ratings, relatively cheap and easy to install. It is supplied on drums in bulk quantities, can be cut to length to meet site requirements and is sufficiently flexible to bend to avoid major obstructions. Such cables are not usually to be found in the house but may be needed for a separate garage or building some distance away.

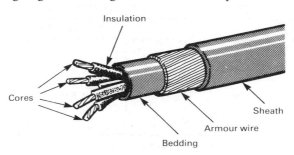

Fig. 48   Armoured cable.

5 Special systems

Harness systems have already been mentioned and they have a special application in housing schemes where the same design and size of 'harness' can be produced in quantities economically in the factory for a large number of similar houses or flats. Almost any form of sheathed cable can be used for such systems.

One rewireable harness system consists of smallbore plastic tubes connected by webs between them, each tube containing a single, loose, p.v.c. insulated wire.

Looped wiring

Conventional diagrams show joints for branch cables as black dots, but to save spending time and labour on joints, slightly more cable is used, and loops are brought into one side of the switch and lamp. Between the other side of the switch and lamp is the switch wire, which should be red. If the lamp is on a fixed bracket or lampholder, the black wire can be looped in and out again, the loop going to one terminal. When pendant fittings are employed, the looping is done at the ceiling-rose instead of at the lampholder, which is connected to the ceiling-rose by a twin flexible cable. This is illustrated in Fig. 49, which shows how three wires enter each switch and the ceiling-rose, but two of them are looped.

When two core sheathed cable is used, it is possible to use a junction box with terminals or cable connectors to enable the lamp to be connected in series with the switch, as shown in Fig. 50. Both red and black wires are fed into a junction box with four entries. Three two core cables are taken out, as indicated in the diagram; one two core cable provides the feed to the next point, the next outlet goes to the ceiling-rose, while the third goes to the switch. The connections within the junction box are made with brass

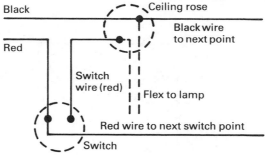

Fig. 49   Looping out to next switch and ceiling-rose

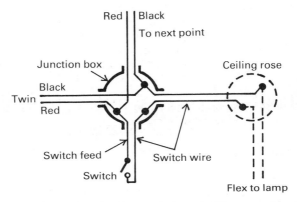

Fig. 50   Twin wiring to junction box, switch and ceiling-rose (earth wire and terminal not shown)

terminals fixed to the base of an insulated box, or with cable connectors having brass terminals encased in an insulating enclosure if the box is of metal. The incoming cable is the fourth two core cable.

Another method, using a three-plate (terminal) ceiling-rose, is shown in Fig. 51, the circuit connections of which are made with pairs of wires or two core cable. This method is more economical in cable and junction boxes.

In all systems of wiring it is most important to connect a single-pole switch in the line or phase lead to the lamp, and never in the neutral or return lead. Regulations prohibit any break in the neutral conductor, except in a linked switch which isolates both poles of the circuit together.

Wiring for various switching arrangements is described in Chapter 7 under Switches and Switching.

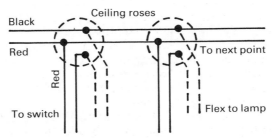

Fig. 51   Three-plate ceiling-rose showing looping out

6

Flexible Cords, Cables, Circuits and Testing

When apparatus is in a fixed position, permanent cabling or 'hard' wiring of a rigid type is normally installed, and for large installations the main current of the switchboard or building may be carried by bus-bars of copper rod or bar fixed on insulators in a metal casing. But in any system provision must be made for connecting portable apparatus in a flexible manner to the permanently-wired outlets. For this purpose flexible leads are required, and these flexible leads are one of the most common points of failure, as they are often misused and are liable to be damaged through carelessness. For household use a flexible lead with a pleasing finish is desirable, while workshop flexibles must be more robust. Flexible leads are colloquially called 'flex', and the conductor is made up of a number of fine wires twisted together and insulated with vulcanised india-rubber (v.i.r.), p.v.c. or some other elastomeric insulation. With v.i.r. and elastomeric compounds, the copper wires are tinned to prevent chemical reaction between the compounds and the copper. It is wise to buy good-quality flexible cords. The style of finish depends on the application and is usually black or white p.v.c. for domestic use, or tough rubber for workshop flex and the size must be selected to suit the current taken by the appliance to be connected. The I.E.E. basic current ratings for flexible cords are given in Table 6 but for larger currents than those given in Table 6, flexible cables from 6 mm^2 to 630 mm^2 are available.

Twin twisted flex consists of two insulated and braided conductors twisted together to form a pair.

Circular flex consists of insulated conductors laid up and twisted together with suitable worming or embedded in an elastomeric compound, the whole being sheathed circular to suit the type of

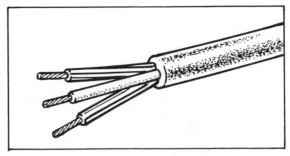

PVC insulated and sheathed flexible
(General purpose domestic use 70°C)

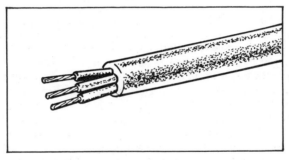

EP Rubber insulated, CSP sheathed flexible
(High temperature domestic use 85°C)

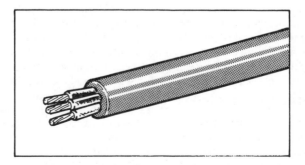

VR insulated Tough Rubber sheathed flexible
(Heavy duty domestic use; oil/grease resistant)

Fig. 52   Flexible cords for domestic appliances

Table 6   Current ratings of flexible cords to B.S. 6500

Conductor Nominal cross-sectional area	Current rating d.c. or single-phase or three-phase a.c.	Maximum permissible weight supportable by twin flexible cord
mm²	A	kg
0·5	3	2
0·75	6	3
1·0	10	5
1·25	13	5
1·5	16	5
2·5	25	5
4·0	32	5

application. Twin twisted flex has been largely superseded by the smoother and cleaner circular form of construction. Fig. 52 illustrates the type of flexibles generally available.

For domestic appliances unkinkable flex is recommended. The rubber insulated conductors are surrounded by a rubber compound and have an overall covering of textile braid. This construction prevents the flex from forming kinks when twisted into a coil and pulled, which often causes fracture of the copper conductors. Most accessories and appliances designed for flexible cord connections have efficient clamps to grip the flexible, as illustrated by the three-pin plugs shown in Figs. 53 and 54. The Regulations require that all portable apparatus, including lamp-standards, must be fitted with suitable types of flexible cord and recommendations are given for various purposes and conditions of use. Unfortunately, this regulation is not always observed with some appliances made abroad. Where exposed to mechanical damage flexible cords must be sheathed with rubber or p.v.c. and, if necessary, armoured.

Three-core flexible

The rule regarding earthing of portable apparatus requires that all exposed metal, with the exception of certain minor isolated parts, must be earthed. For this purpose a three-core flex is used, which consists of three insulated conductors. The live core should be coloured brown and must go to the switch, if one is fitted on the appliance, blue core is the neutral; and the green/yellow striped core goes to the earthing terminals of the appliance and the three-

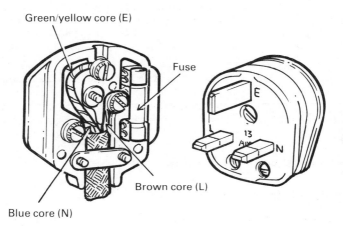

Fig. 53   Three-pin plug (flat pins) and flexible lead
(*M.K. Electric Ltd.*)

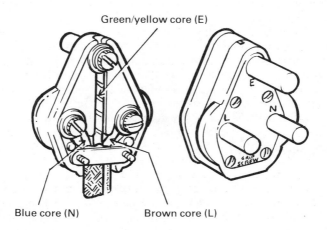

Fig. 54   Three-pin plug (round pins) and flexible lead
(*M.K. Electric Ltd.*)

pin plug. The earthing-pin is the larger of the three pins and is generally arranged to be at the top of the socket outlet. The method of attachment and connections are shown in Figs. 53 and 54.

Before 1970 the established colours of cores in flexible cords were red for the 'live' conductor, black for the neutral conductor and green for the earthing conductor, and some such flexibles may still be in use, but international standards of colouring have been adopted which avoid the confusion and danger of using imported electrical appliances with flexibles having different-coloured cores. The correct standard colours are now brown for the 'live' conductor, blue for the neutral conductor and green/yellow stripes for the earthing conductor in three-core flexibles.

Choice of flexible cords

Sheathed flexible cord with armouring where necessary is employed for trailing cables, in damp positions and where good mechanical protection is essential. Tough rubber sheaths (t.r.s.) are the most flexible and are least affected by greases and oils; p.v.c. sheaths, however, are softened by oil- and grease but are also easily cleaned. Special flexibles for very arduous duties and use where subject to abrasion are protected by a wire braid armour over the sheathing. Such flexibles find application in many farm buildings.

All electrical equipment generates heat when in operation, and it is a design problem to dissipate the heat without excessive temperatures developing inside or on the outside of the appliance whether it is a luminaire or a cooker. The flexible lead to a luminaire or a portable heating appliance passes through hot surfaces and is in hot surroundings where it enters the casing or enclosure; heat damages rubber and thermoplastic insulation and sheathing; all lighting pendant flexibles are subject to excessive temperatures where they emerge from the lampholders; heat-resistant flexibles must therefore be used for all such applications. The common heat-resistant insulating materials for flexibles – in ascending order of heat-resisting quality – are: Butyl or Ethylene Propylene Rubber and heat-resisting p.v.c., varnished p.t.p. fabric, silicone rubber and varnished glass fibre.

The different uses of portable appliances create mechanical stresses in the flexible and its construction must take account of these. While there is negligible stress and flexing in a simple plain light pendant, considerable stress and flexing may occur in a rise-and-fall pendant. The use of gardening appliances also leads to great stress in their usually extra long flexible leads; these leads are also subjected to damp. Some of the types of flexibles designated for various purposes in the Regulations are as follows:

Indoors with low mechanical stress and occasional outdoor use	60°C rubber-insulated and sheathed.
Light duty indoors with dry conditions	light p.v.c.-insulated and sheathed.
Medium duty indoors with damp situations, cooking and heating appliances (but not hot parts), outdoor use	ordinary p.v.c.-insulated and sheathed.
Any use including hot situations such as night storage and immersion heaters	85°C rubber-insulated and H.O.F.R. (heat and oil resisting and flame retardant) sheathed.
In high ambient temperatures and in or on luminaires	150°C rubber-insulated and braided.
Luminaires only	185°C glass-fibre-insulated single-core, twisted twin and 3-core.
Luminaires and dry situations with high ambient temperatures not subject to abrasion or undue flexing	185°C glass-fibre-insulated, braided, circular.

Short lengths of sleeving in the above high temperature materials are available to protect the last few inches of an otherwise unsuitable cable.

Other flexibles and parallel twin and twisted twin, non-sheathed, for luminaries as permitted by B.S. 4533 are also included.

Sizes of conductors

The current ratings and consequent sizes of cables and wiring conductors depend on whether the cables are to be run singly or bunched together, and whether they are supported in air or enclosed in a conduit or trough, or buried in walls or the ground. Different current ratings are tabulated for these conditions as well as correction factors for ambient temperature; ratings for two typical conditions are given in Table 7 (page 85).

Other considerations that affect the size to be used are:

(a) The voltage drop, due to the resistance of the conductor and the current it has to carry.

(b) The heat-dissipating qualities of the type of cable insulation to be used and the ambient temperature.

(c) The minimum size of wire to be permitted by the circuit design.

Table 7 Current ratings for loaded single-core p.v.c.-insulated cables (copper conductors) run bunched, and enclosed in conduit or trunking, with ambient temperature of 30°C and where over-current protection is provided by domestic type h.r.c. (high rupturing capacity) cartridge fuses or miniature circuit-breakers. Where semi-enclosed (rewireable) fuses are provided for protection these current ratings must be reduced by 27%.

Conductor	Two cables d.c. or single phase a.c.		Three or four cables three phase a.c.	
Nominal section	Current rating	Voltage drop per ampere per metre run	Current rating	Voltage drop per ampere per metre run
mm²	*A*	*mV*	*A*	*mV*
1	14	44	12	38
1·5	17	29	16	25
2·5	24	18	21	15
4	32	11	28	9·5
6	41	7·3	36	6·4
10	57	4·4	50	3·8
16	76	2·8	68	2·4
25	101	1·8	89	1·6
35	125	1·3	110	1·1

Voltage drop in cables

The effect of incorrect voltage has been dealt with in Chapter 2 (see p. 36), but here we are concerned with the effect of the cable sizes chosen on the voltage available at the terminals of the lamp or other apparatus.

The rule for the maximum allowable voltage drop is 2½% of the supply voltage.

With the standard voltage of 240 this is:

$$\frac{2 \cdot 5}{100} \times 240 = 6 \, \text{V}.$$

On a 100 V circuit it is 2·5 V, and as the current per watt is greater much larger sizes of cable must be used to keep within the permissible voltage drop with such a low voltage supply as this.

Of this figure, about 2% drop should be spread over the main cables and those forming the final circuits from the distribution boards in about equal proportions for economy in wiring. The remaining ½% is left for the voltage drop at contacts of distribution boards, switches, etc. In long runs of mains it is desirable to keep the cable sizes larger than those required by the allowable voltage drop. This allows for any future extensions and growth of load, and also minimises the effect of lower supply voltage, which may occur at times of peak load on the supply network. With large loads such as 3 kW electric fires or water heaters at the end of a long run, the relatively large current taken by such apparatus affects the voltage of all the other circuits, so that some liberality in cable sizes is justified. Table 7 also gives values of voltage drop for various cables used in house wiring. For more complete lists, reference should be made to tables in the I.E.E. Regulations.

The rated currents given apply at an ambient air temperature of 30°C (86°F). Above this temperature, or with larger groups of conductors, the current rating is reduced by factors specified in the Regulations. P.v.c. insulated, wire armoured and sheathed cables have better heat-dissipating qualities and can carry much greater currents than those given in Table 7. Sheathed p.v.c. cables embedded direct in plaster have around 10 per cent higher current ratings than they would if carried in conduits due to the better heat dissipation.

A further modification in the current-carrying capacities of cables depends on the kind of over-current circuit protection provided for the cables. Table 7 gives values for what is considered to be the latest standard of domestic circuit protection, but there will be many installations with semi-enclosed (rewireable) fuses for some time to come on account of their cheapness. With the form of over-current protection provided by high rupturing capacity cartridge fuses and miniature circuit-breakers, the current-carrying capacities of the cables are greater because the protective device will operate quickly (within 4 hours with 50% excess current), whereas rewireable fuses will take longer to operate, allow a dangerous fault current to persist for too long and cause damage by overheating the cable. Apart from the obvious advantages of the higher standard of protection the question of whether the higher cost is balanced by the saving in using smaller cables depends on the particular requirements of the installation design.

Choice of cable size

It will be evident that the maximum permissible current that any cable can carry will depend on a number of factors that must be applied to the basic figures in tables. These factors are given in appropriate tables for various types of cable in the I.E.E. Regulations, and they cover type of protection, grouping, conditions of installation and ambient temperature. The current required can be estimated and the most suitable size of cable selected. For small house installations the smallest appropriate cables are usually adequate for mains, lighting and socket outlet circuits without regard for voltage drop or other factors involved, but where there are long runs the voltage drop should be checked to see that it is satisfactory. The smallest size of conductor for final circuits is 1 mm^2, rated at 14 A, and with flexible cords the smallest size is 0·5 mm^2, rated at 3 A basically in the Regulation-tables.

Before metrication in the UK, the two available single-strand cables had been found so stiff and difficult to draw into conduit, and so susceptible to fracture that the smallest three-strand and more flexible cable was almost always used instead, although slightly larger in size. Metrication, however, has resulted in the introduction of more sizes of single-strand cable which are all larger than the previous two and even more difficult to handle. Although this is an advantage for sheathed surface wiring, it does increase the labour involved in drawing wires into conduits.

Voltage drop calculations

From Ohm's Law we know that the voltage drop = $I \times R$, where I is the current in ampères and R is the resistance in ohms. From the load in watts (P), the current (I) can be obtained by dividing by the voltage (V), i.e. $I = P/V$. The resistance per kilometre of various cables is given in wire tables, so the 'voltage drop' can be calculated. All the 'drops' of conductors in series are then added together from the service up to the apparatus concerned and should not exceed 2½% of the declared voltage.

EXAMPLE 10.   The load in a house is 8 kW and the supply voltage is 240. If the distance from the incoming supply to the distribution fuseboard is 10 m, what will be the voltage drop and the size of the two single-core main cables?

$$\text{Current } I = \frac{P}{V} = \frac{8 \times 1000}{240} = 33 \cdot 3 \text{ A}.$$

Referring to Table 7, and assuming no correction factors are necessary for temperature, grouping or form of protection, two single-core 6 mm² cables in conduit or casing will carry 41 A. The total length of the cable, flow and return is $2 \times 10 = 20$ m, and its resistance per kilometre is 2·97 Ω. Thus the total resistance (R) is $^{20}/_{1000} \times 2·97 = 0·059$ Ω. Hence the volt drop is $Vd = IR = 33·3 \times 0·059 = 1·96$ V.

This is well within 2½% of the supply voltage (6 V), but there will also be some voltage drop in the final circuit cables.

With a number of lamps and other apparatus, the total load is distributed over the length of final circuit cable from the distribution board. The farther away a point is from the distribution board, the less is the current but the greater is the cable resistance. Thus the total load can be considered as concentrated at some point along the final circuit cable. This 'load-centre' may be the geometrical mid-point of the group, or it can be found in the same way as a centre of gravity (or centroid) in applied mechanics. With a well-distributed load it can be assumed to be halfway along the final-circuit route. To find the percentage voltage drop measure the distance from the distribution board to the load centre, add up the maximum load on the branch circuit in watts and work out:

$$\text{Percentage voltage drop} = \frac{PRL}{10\ V^2}$$

where P = load on the circuit in watts,

$\quad\quad R$ = resistance per kilometre of the conductor used,

$\quad\quad L$ = length of circuit including return, i.e. twice the route length in metres,

$\quad\quad V$ = declared supply voltage.

EXAMPLE 11.   The lamp load on a circuit is 650 W, and the route length is 30 m. If the cable has a resistance of 17·1 Ω per kilometre, what is the percentage voltage drop on a 240 V supply?

Using the formula above:

$$\text{Percentage voltage drop} = \frac{650 \times 17·1 \times (2 \times 30)}{10 \times (240 \times 240)} = 1·16\%$$

so the actual voltage drop is $1·16/100 \times 240 = 2·78$ V. This, added to the voltage drop in the main cables (1·96 V), makes a total of 4·74 V, which is well within the permitted 2½%, so the cable size is satisfactory.

There will be a slight difference between the values of voltage drop calculated from the conductor resistances and those worked out from the volts drop per ampère per metre run given in Table 7. This is because the former are standard values for cable conductors at 20°C, whereas the latter are values for cables in an ambient temperature of 30°C carrying the rated current. Complicated temperature correction factors are involved to account for the differences. However, for all practical purposes it is simpler and satisfactory to use the voltage drop figures in the cable current tables, on which Table 8 is based for a group of two cables. This shows the maximum length of run in metres for 1% voltage drop on a 240 V system with various load currents for five small cable sizes.

Table 8 Maximum length of run, in metres, for 1% voltage drop on a 240 V supply (two cables). Figures are not given above the stepped lines because the current would exceed the maximum permitted for the cable.

Note: These figures are approximate as the voltage drop basis in the cable current-carrying capacity tables is only correct for the full rated current of the cable; corrections for temperature at different loads, which affects the conductor resistance, would be necessary for accurate figures.

Load in W		6000	5000	4000	3000	2000	1500	1250	1000	750	600	500
Current in A		25	20·8	16·7	12·5	8·33	6·25	5·2	4·16	3·12	2·5	2·08
Conductor												
Size mm²	Rated current (A)											
1	14				4·8	7·2	9·6	11·6	14·4	18	24	29
1·5	17			5·3	7·1	10·7	14·2	17	21	28	36	43
2·5	24		7·2	9	12	18	24	29	36	48	60	72
4	32	9·6	11·5	14·4	19	29	38	46	58	77	96	116
6	41	14·1	17	21	28	42	56	68	84	112	140	169

Effect of a heavy load on the end of a sub-circuit

EXAMPLE 12. Fig. 55(a) shows a pair of main cables, each with a resistance 0·05 Ω and a final-circuit with one 100 W lamp halfway along it and another 100 W lamp at the end. The resistance of each final-circuit wire is 1 Ω. With the first 100 W lamp switched on to, say, a 110 V supply, a current of $^{100}/_{110} = 0.91$ A flows along the main cables and half the length of the final circuit. The total resistance is $0.05 + 0.5 + 0.5 + 0.05 = 1.1$ Ω, so the IR drop = $0.91 \times 1.1 = 1$ V.

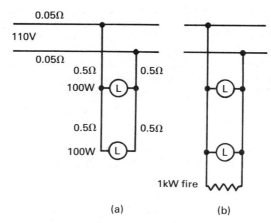

Fig. 55 Illustrating voltage drop and heavy load on long sub-circuit (circuit fuses and switches not shown for simplicity, and 110 V chosen to accentuate the effect)

Thus the voltage at the first lamp will be 109 V. When the second lamp, at the end, is switched on, the total current will be 1·82 A from the mains. The voltage drop is 1·82 × 1·1 = 2 V at the first lamp plus 0·91 × 1 = 0·91 V at the second lamp. Thus the voltage across the second lamp is 107·09, say 107 V. This illustrates how the lights may be less brilliant as one proceeds along a final circuit. Now, suppose the occupier thinks he would like to have a 1 kW fire in the same room as the last lamp and connects the fire to the end of the final circuit (see Fig. 55(b)). This fire will take $^{1000}/_{110} = 9·1$ A, which is added to the existing lighting load and was never contemplated when the wiring was put in. Now let us work out the voltage across each lamp.

The total current from the mains is 9·1 + 1·82 = 10·92 A. Drop to first lamp = 10·92 × 1·1 = 12·01, say 12 V, so the first lamp has only 98 V across it, or 89% of its proper voltage. In the remaining half of the final circuit the current is 9·1 + 0·91, say 10 A, and the further drop in voltage is 10 × 1 = 10 V; so the second lamp has only 98 − 10 = 88 V across it, which is 80% of its proper voltage, and thus it will burn dimly. These abnormally high resistance figures have been chosen deliberately to bring out the effect of a heavy load on the end of a final circuit with lamps, which is not good practice. The correct way is to run a separate circuit for the electric fire. This is shown in Fig. 56, in which the resistance of the wires to the fire are each taken as 0·125 Ω. We will now calculate the currents and voltages across the lamps. The main cable still carries 10·92 A and the drop is 1·09 V. The drop in the wires to the fire is 9·1 A × 0·25 = 2·27 V, so the fire is on 108·9 − 2·27 = 106·6 V, which is not too bad. The voltage drop to the first lamp is 1·82 V, giving 110 − 1·09 − 1·82 = 107 V, while the second lamp is on 107 − 0·91 = 106 V (approx.). This compares favourably with the figures of 98 V and 88 V respectively in the former case.

Standard ring and radial circuits

From the above example it can be seen that power and lighting circuits should be run separately. In radial final circuits it would be sufficient to choose the cables for the circuits that supply electric fires and other apparatus on a current-carrying basis, as the current taken is separate from the lighting circuits and the voltage drop is less important. For instance, in considering a very simple radial circuit for one point, a 2 kW fire on 110 V takes 18·2 A, so that 2·5 mm^2 cable could be used. The same size of fire for 240 V takes 8·3 A, so 1 mm^2 cable would be large enough. This would also take a larger fire, up to 3 kW, and if a 13 A socket-outlet was provided, the 1 mm^2 cable to carry 14 A would still be large enough. For apparatus taking up to 1 kW on 200–250 V, a 5 A plug and socket could be used. For larger currents between 5 A and 15 A, there is the choice of 13 A or 15 A plug size, e.g. 3 kW at 240 V takes 12·5 A; so a 13 A or 15 A outlet could be employed.

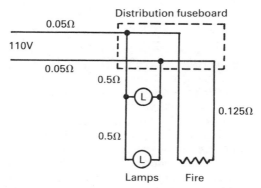

Fig. 56 Correct way to connect a fire as a separate circuit (circuit fuses and switches not shown for simplicity)

In modern UK practice the 13 A socket is the standard all-purpose domestic socket outlet for a.c., and, because of the diversity of usage, ring and radial final circuits for 13 A socket outlets may supply an unlimited number of points (with the exception of fixed water-heating points and the fixed heaters of a comprehensive space heating system which should have separate circuits) in specified areas with particular over-current protection and cable sizes as follows:

1. Area up to 100 m^2; with (i) 30 A or 32 A fuses or any other type of protection and **ring** circuits: or (ii) area up to 20 m^2 with 20 A rated protection (any kind) and **radial** circuits of:
   (a) 2·5 mm^2 copper wire and rubber or p.v.c. insulation; or
   (b) 4 mm^2 copper-clad aluminium wire with p.v.c. insulation; or
   (c) 1·5 mm^2 copper wire with mineral insulation.

2. Area up to 50 m^2; with 30 A or 32 A rated cartridge fuse or circuit breaker protection and **radial** circuits of:
   (a) 4 mm^2 copper wire and rubber or p.v.c. insulation; or
   (b) 6 mm^2 copper-clad aluminium wire with p.v.c. insulation; or
   (c) 2·5 mm^2 copper wire with mineral insulation.

3. Non-fused spurs are connected at joint boxes or at socket outlet terminals or at the distribution board and must have the same size of cable as the main circuit and feed not more than one single or one twin socket outlet or one fixed appliance; but the number of spurs must not exceed the total number of points directly connected to the circuit. The number of fused spurs is unlimited and they are connected through a fused connection unit, the rating of which must not exceed 13 A. For socket outlet spurs the cable must not be smaller than 1·5 mm^2, 2·5 mm^2 or 1 mm^2 respectively for the types of cable in (a), (b) and (c) above.

When ring circuit cables are bunched together or the ambient temperature is over 30°C the cable size must be increased so that its current-carrying capacity is two-thirds the rating of the circuit protective device in the previous section 1; and in the case of radial circuits, it must be increased to the full rating of the protective device in sections 1 and 2; using the appropriate correction factors. Where the number of socket outlets in a circuit is counted, each outlet in a twin or multiple group at a single wiring point must be counted separately. Kitchens may need a separate circuit. In domestic installations with a.c. supplies the great diversity allowed in the provision of an ample number of 13 A socket outlets for general purposes throughout the house, with several in each room, make the ring circuit a very economical arrangement of wiring, with a capacity of only a small fraction of that represented by the number of socket outlets connected to it. The wiring forms a ring with both ends connected to the circuit protective device, the wiring being looped into each point. It is often found that two ring circuits, one for the ground floor and one for the first floor, are adequate for the majority of medium-sized houses; and for the small house a single ring circuit for ground floor with vertical spurs to first floor points may be sufficient for normal requirements. The 30 A or 32 A circuit protective device limits the total loading to 7·2 kW or 7·7 kW respectively (at 240 V) and the small cartridge fuses of 3 A or 13 A capacity in the plugs protect each standard-lamp or other appliance connected to the ring circuit independently. The ring circuit diagram is shown in Fig. 57.

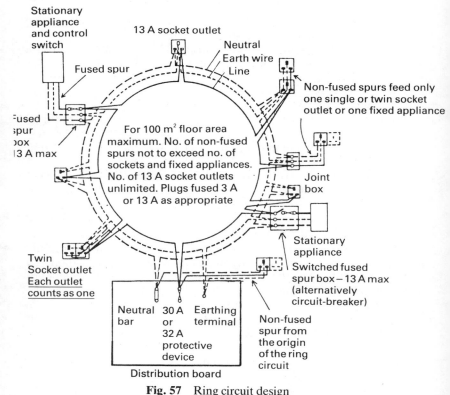

Fig. 57   Ring circuit design

(*A.S.E.E.*)

Lighting circuits

The current rating of lighting final circuits is based on the connected load, but not less than 100 W per lampholder; discharge lamp current is taken as 1.8 times the wattage divided by the voltage; and all the other small points such as electric clocks, shaver units, bell transformers, etc., with ratings less than 5 VA can be neglected. In domestic premises either 1 mm or 1·5 mm circuit cable is usually large enough, and the latter will, of course, have less voltage drop. Any 5 A or 15 A sockets should be taken at their full value, and 2 A sockets can be taken at ½ A since, in general, these are only used for lighting purposes.

Other circuit ratings and diversity of installation loads

In medium-sized houses it is not usual for more than half the lighting to be on at the same time, though two-thirds is taken in calculating the maximum demand of the premises. This should not lead to a reduction in cable size of the lighting circuits in the house as this degree of diversity only applies in calculating the loading of the main cables for the whole installation.

To determine the rating of a final circuit and the installation load for stationary cooking appliances, the total current is assessed by taking the first 10 A of the total rated current of the appliances, plus 30% of the remainder, plus 5 A if the cooker control unit incorporates a socket outlet.

The current rating of all other circuits is taken as the sum of all the equipment and outlet ratings and applying a diversity factor if necessary; stationary appliances are rated at their normal current.

For the installation load of heating and power appliances, the full load up to 10 A is taken, plus 50% of any load in excess of this value. Water heaters (thermostatically controlled), floor warming and thermal storage space heating installations are taken at their full rated currents bearing in mind that off-peak thermal storage loads generally do not coincide with the other loads in the house. The loading of 13 A general-purpose standard socket outlet ring and radial circuits can be taken as the full rating of the largest fuse or circuit-breaker rating of the individual circuits (which is 30 A or 32 A for a ring circuit), plus 40% of the sum of the fuse or c.b. ratings of the other circuits (diversity is already allowed for in the circuit design). For other socket outlets and stationary appliances, a similar method applies, i.e. the rating of the largest point, plus 40% of the sum of all remaining point ratings. If there are more than two instantaneous type water heaters an allowance of 25% full load of the remainder is made.

All sections of the installation load, so calculated, are added together to obtain the maximum load that will be demanded from the supply undertaking and carried by the main cables and switch-gear. In larger buildings and institutions, different proportions are employed. Except for cookers, these values should not be applied in assessing the loading of distribution boards as they only represent the diversity of demand to be expected in a complete installation.

Installation testing

When a new or extended installation has been completed the installer should provide adequate information in the form of dia-grams, charts or schedules of points, especially showing the fusing and circuit arrangements so that future maintenance and repairs can be undertaken with safety. All equipment should be checked to ensure compliance with British Standards, that it is in good con-dition and that the installation is in accordance with Regulations.

Testing should be carried out in a logical sequence to ensure that all items are covered and that where one test depends on the satisfactory completion of another test they are carried out in the correct order. For instance, it is of little use testing insulation resistance if it is not known whether all the circuit conductors are continuous and whether switches and outlets are correctly connected.

The required sequence of testing is given in the Regulations which should be consulted for detailed methods of testing but the following covers the principal items of the tests and in the order stated.

Sequence of testing:
(a) Continuity of ring final circuit conductors
(b) Continuity of protective conductors, including main and supplementary equipotential bonding
(c) Earth electrode resistance
(d) Insulation resistance
(e) Insulation of site-built assemblies
(f) Protection by electrical separation
(g) Protection by barriers or enclosures provided during erection
(h) Insulation of non-conducting floors and walls
(i) Polarity
(j) Earth fault loop impedance
(k) Operation of residual current devices

This list may appear daunting but is intended to cover protection against all eventualities. For electricity in the house, the individual tests are simple to carry out but require some specialist test equipment so it is recommended that testing be carried out only by an experienced Electrician or Engineer.

Insulation tests are also usually carried out by the supply undertaking at the intake position before the installation is allowed to be connected to the service cable. They are a check on the overall safety of the installation, but such tests, even if the test readings are quite satisfactory, are not in themselves a guarantee of the safety of the installation, for which further tests are necessary. It is important to note that no tests other than earth-loop impedance tests should be made on an installation unless the supply is disconnected and all wiring is 'dead'.

It is recommended that the installation be periodically inspected and tested at intervals of not more than five years, and an inspection and test certificate should be issued by a competent person as described in the I.E.E. Regulations.

Continuity of final circuit conductors
Methods of testing the resistance of circuit conductors are shown in Fig. 58 (page 97). In the case of a ring circuit the resistance is taken between the ring terminal at the distribution board and the corresponding terminal of the socket outlet nearest the midpoint of the ring, using a suitable lead of known resistance which is subtracted from the ohmmeter reading. The result must be approximately a quarter of the resistance of the ring circuit cable taken separately before connection. The resistance of each of the circuit and protective earthing conductors in a radial circuit is found by connecting it in series with one of the other circuit conductors, testing the loop and subtracting the known resistance of the circuit conductor or such other cable used for this purpose.

Continuity of protective conductors
A test for continuity of steel conduit or other steel enclosure and of earthing conductors and equipotential bonding is made by applying a d.c. or a.c. voltage that need not exceed 50 V with 1.5 times the rated current of the circuit but which need not exceed 25 A. For protective and bonding cable conductors a d.c. ohmmeter can be used for the resistance test.

Earth electrode resistance

In the average town house, the Electricity Supply Authority will provide a point to which all protective conductors and equipotential

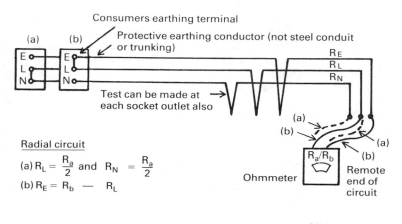

Radial circuit

(a) $R_L = \dfrac{R_a}{2}$ and $R_N = \dfrac{R_a}{2}$

(b) $R_E = R_b - R_L$

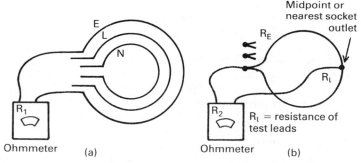

Ring circuit

(a) Resistance of each conductor, in turn = R_1
(before connecting to distribution board)

(b) (after connecting to distribution board) = $R_E = R_2 - R_l = \dfrac{R_l}{4}$ (approx.)

(with ring connection complete)

Fig. 58 Testing resistance of conductors in ring and radial circuits

bonds can be made. This will usually be an extension of the metal sheath or armouring of the main incoming supply cable or a separate earthing conductor connected to earth outside the house. In the case of installations in houses remote from public supply, an earth electrode system should be provided. This involves running an earthing conductor from the main earth terminal of the installation to an earth electrode, usually a metal plate or a series of metal rods, buried or driven into the ground outside the house as well as earthing the neutral point of the supply at the transformer or the local generator. The design of the system of earthing depends so much on the electrical characteristics of the ground in the area and other local conditions that it is not possible to give more information than this in this book.

The electrical resistance of the earth electrode and the resistance of the earth path from the electrode to the supply neutral point can be measured directly by use of a suitable specialised instrument, a trade name for which is an *Earth Megger*.

Insulation resistance

The insulation resistance of the system depends but slightly on the grade of cable used, provided that there are no breaks in the insulation covering the wires. Most of the leakage current occurs at cable ends, switches, distribution boards and points where the connections are made. The total insulation resistance decreases with increase of length of cable runs and with more outlets. The rule is that the insulation resistance of the wiring shall not be less than 1 MΩ between line and neutral conductors, and between conductors and earth excepting, of course, the neutral conductor where this is earthed and also used for protective earthing in the installation. The insulation to earth of fixed apparatus that is disconnected for the tests may be as low as 0·5 MΩ.

To test the wiring an ohmmeter is used. A trade name for such an instrument is the *Megger*; this instrument consists of a hand-driven generator which supplies a constant d.c. voltage to the installation under test. The test voltage required is 500, or twice the working voltage of the installation, in the case of domestic premises. There are two terminals on the *Megger*, one labelled earth and the other line. When the handle is turned, a constant speed is ensured by means of a slipping clutch. The pointer on the instrument moves over a scale of ohms, which varies from infinity to zero. For the larger values, megohms, or millions of ohms, are marked on the scale. Modern types of insulation and continuity testing instruments use electronic internal circuitry and small dry batteries for the current supply but can still produce the test voltage mentioned above.

With a perfect insulator, the pointer goes to infinity, while with a perfect conductor, i.e. perfect continuity, the pointer goes to zero. Thus this instrument can be used for testing insulation resistance or continuity of conductors and the bonding of extraneous metalwork liable to become live.

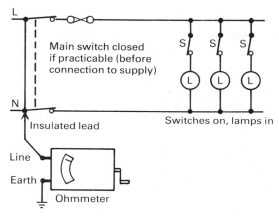

Fig. 59 Insulation test to earth before supply is connected

Two tests of insulation are performed, one 'to earth' and the other 'between conductors'. Fig. 59 illustrates the insulation test to earth for a radial circuit, which is made with all fuses in, switches on and all poles or phase conductors connected together (not applicable to installations with the neutral conductor and the protective earthing conductor bonded together or being common). The line terminal of the ohmmeter is connected to one conductor, while the earth terminal is connected to a good earth, such as a metal water pipe that enters the ground (not a gas pipe) or the supply authorities' earthing terminal. The second test checks the insulation resistance between the two conductors, during which the lamps are removed and all apparatus is disconnected, as shown in Fig. 60.

Insulation by site-built assemblies, electrical segregation, barriers, enclosures and non-conducting surfaces
In exceptional cases, it may be necessary to construct on the site of the electrical installation some enclosures, barriers, or other means of insulation or segregation of the electrical systems from the normal environment and these must be visually and electrically tested to ensure personal safety, insulation, earthing, bonding and compliance with Regulations. In addition, in some specialist areas, reliance is placed on conducting or non-conducting flooring, walls or structures and these must be verified but it is not likely that cases of this kind will be found in a typical house.

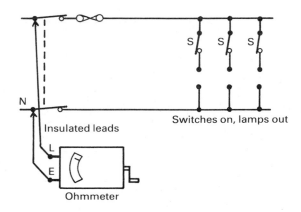

Fig. 60   Insulation tests between poles before supply is connected

Polarity testing

Testing for polarity, to ensure that all switches are connected in the live phase conductor and not in the neutral conductor, is very important. This test can be carried out with an ohmmeter or with a simple bell and dry battery arrangement. After energisation of the electrical installation the polarity can again be checked using instruments or neon indicators to show the circuit conductor with the highest voltage, in which the switch should always be inserted.

Earth fault loop impedance

In the event of any insulation becoming defective so that a current carrying conductor comes in contact with the metal casing, the earth connection provides a low resistance path to earth and the large fault current will open the circuit by operating the fuses or circuit-breaker within a reasonably safe period, thereby preventing a dangerous electric shock to anyone touching the defective apparatus. About 50 V d.c. or 30 V a.c. is the maximum pressure that a human being can safely withstand, but about half this value is the safe limit for animals. Alternating current is more dangerous than direct current, but in any case the effects depend on the physical condition of and the positions of the contact points on the person receiving an electric shock. Therefore it is necessary to correlate prospective shock voltage with the kind of circuit protective device to be used and the time it will take to operate as well as with the impedance of the earth fault loop, in order to arrive at a safe limiting value of the protective conductor impedance. This is taken into account in the I.E.E. Regulations tables of earth fault loop impedance and protective conductor impedance.

With fixed equipment not having any exposed hand-held conductive parts the disconnection time for circuit protective devices is five seconds since the physical conditions limit the risks involved. But if there is equipment with such exposed parts or the circuits supply socket outlets (as they do in a house) the operating period is reduced to 0·4 of a second because hand-held appliances are much more dangerous if live. The calculations necessary to work out the limiting value of the installation protective earthing conductor are complicated, but tables in the Regulations simplify the problem by giving limiting values of impedances of protective conductor and total earth loop impedance, for the various types of fuse and circuit-breaker that can be used. From these references the size of the protective conductor can be deduced or checked, together with its current-carrying capacity. The limiting value must, of course, apply to the full length of the circuits from the farthest points.

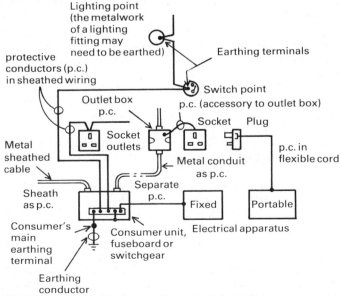

Fig. 61 Methods of achieving earth continuity with protective conductors

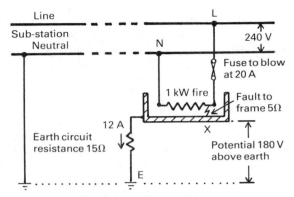

Fig. 62 Earth loop circuit of high resistance

Fig. 61 shows the various ways continuity of the protective conductor is achieved in an installation and the following example will show an earthing connection that is worse than useless with simple calculations to illustrate the basic principle involved.

EXAMPLE 13.   The end of a 1 kW electric fire element adjacent to the 'live' conductor develops a fault to the metal casing of the fire. The resistance of the fault is 5 Ω and that of the earthing circuit 15 Ω. The supply is 240 V and the fuse fitted will blow at 20 A. Is this apparatus safe and properly protected?

The circuit diagram is given in Fig. 62, where the neutral conductor is shown earthed at the sub-station and the earth connection or electrode is denoted by E. The resistance from L to E via the fault X is 20 Ω, so the fault current is $^{240}/_{20} = 12$ A. This circuit is in parallel with the fire element, which takes a current of $^{1000}/_{240} = 4 \cdot 2$ A approximately. Therefore, the total current is $16 \cdot 2$ A, which will not blow the fuse or overload the cable, assuming this is $1 \cdot 5$ mm^2 to carry 17 A. But there is 240 V from L to E, as the resistance between E and the sub-station earth is taken to be negligible in this case, and three-quarters of this potential difference (in proportion to the resistance) is between the points X and E, so that any person who was 'earthed', say by standing on a damp concrete floor, and who touched the frame would be liable to an electric shock at 180 V; the prospective shock voltage in this case is, of course, 240 V; damaged or bad contacts at the earthing terminals could cause such a high resistance and shock voltage. The calculation can be done another way: the potential *above* earth at point X is the fault current, 12 A, multiplied by the earthing circuit resistance of 15 Ω, namely 180 V.

It will be seen that this apparatus is most unsafe and that the fuse gives no protection against such a fault. Suppose the fault develops into a direct earth of zero resistance and the fault current then increases to $^{240}/_{15} = 16$ A, so that the total current through the fuse is now $16 + 4\cdot2 = 20\cdot2$ A. The fuse will heat up, and may not blow at once and clear the circuit; but the leakage current will raise the potential of the frame to 240 V – the mains voltage – which is obvious from Fig. 62. Thus the resistance of the earth circuit must be well below 15 Ω for the fuse to blow promptly.

Whether the protective conductor is a separate copper conductor, metal sheathing or conduit, its resistance must be low enough for the fault current to operate the protective device within prescribed time limits for the type of circuit or equipment involved, hence the regulation to limit this resistance.

Although for the purpose of Example 13 the ground resistance to the electricity sub-station earth was taken as negligible, it is not always so. In large towns it is generally very low, but in some country districts it may be appreciable and may seriously affect the usefulness of protective devices. Therefore, it is important to ascertain the earth-loop impedance. Impedance is the same as resistance in d.c. circuits, but with a.c. circuits it also takes into account any voltage reaction that may be present due to inductive effects (see pp. 10, 220). The earth-loop impedance includes the earth resistance itself, as well as the impedance of the conductors in the sub-station, mains and house wiring, and the protective earthing conductors of the installation.

The earth fault loop impedance is tested by passing a current through this circuit by means of special testing equipment, which can be plugged into a live socket outlet, and the reading in ohms must be within suitable limits for the protective devices installed. Some figures extracted from tables in the Regulations are shown in Table 9. In this table the maximum earth fault loop impedances for various types of protective device enable disconnection within 0.4 of a second in the case of socket outlet circuits and 5 seconds in the case of circuits for fixed equipment, assuming the impedance of the earth fault itself is negligible. The Type 3 M.C.B. requires lower loop impedances because it takes about 2·5 times the current of Type 1 to operate in the same time and limit the shock voltage to the same value for safety.

The result of these limitations is to ensure that, in the event of an earth fault, the earth-leakage current will be high enough to blow the fuse promptly and so avoid a dangerous shock voltage existing on exposed earthed metalwork, e.g. in a socket outlet circuit with a

Table 9 Maximum earth-fault loop impedances for overcurrent protective devices (230–250V) (a) For socket outlet circuits. (b) For fixed equipment circuits.

Current rating of protective device	Earth-Loop Impedance, ohms					
	Fuses				Miniature Circuit-breakers B.S.3871	
	Semi-enclosed B.S.3036		Cartridge (Domestic) B.S.1361			
					Type 1	Type 3
A	(a)	(b)	(a)	(b)	(a)	(b)
5	9·6	20	11·4	17	12	4·8
10	—	—	—	—	6	2·4
13*	—	—	2·5	—	—	—
15	2·7	5·6	3·4	5·3	4	1·6
20	1·8	4	1·8	2·9	3	1·2
30	1·1	3·2	1·2	2	2	0·8
45	0·6	1·6	0·6	1	—	—
50	—	1·2	—	0·6	1·2	0·48
80	—	—	—	0·48	—	—
100	—	0·55	—	0·28	—	—

* Used in 13A plugs

20 A rewireable fuse, and with an overall earth-loop impedance of less than 1·8 Ω we can rest assured that the current will blow the fuse within 0·4 of a second and the shock voltage will be well below the danger limit when the fuse operates, so the installation will be safe to use.

Operation of residual current devices

Where the earth-loop impedance is greater than the permitted figures and cannot be reduced sufficiently by increasing the size of the protective earthing conductor or by using sufficiently sensitive over-current devices, then protective devices of greater sensitivity are necessary. These are the current-operated earth-leakage circuit-breakers, which do not provide over-current protection and must be used in addition to the normal over-current protection devices, such as the fuse or miniature circuit-breaker. They effectively prevent a shock voltage on exposed metalwork from exceeding 50 V, which is considered safe enough.

An earth-leakage trip may be combined with an overload device, with which it is mechanically interlocked, and this combination can take the place of the main switch and fuse at the minimum of additional cost for the extra protection. But overload (over-current) and earth-leakage protective devices are very often separate units so that they can be isolated more easily by the main switch for maintenance purposes.

The residual current circuit-breaker

An automatic Residual Current Circuit-Breaker (RCCB), originally called an Earth Leakage Circuit Breaker (ELCB) consists of a lightly set circuit-breaker operated by a trip coil which carries the earth-leakage (residual) current.

This is called a Residual Current Device (RCD) or Circuit-Breaker (RCCB) because the line and neutral conductors are linked with an operating coil in a magnetic circuit, so that earth leakage causes unbalanced currents to flow through line and neutral conductors thus energising the operating coil of the circuit-breaker; with no leakage, the currents are equal and the coil remains inoperative. The connection diagram for a typical RCCB is shown in Fig. 63.

The RCCB is commonly used where good earthing is difficult and the earth-loop impedance is too high for any form of over-current device to afford adequate protection. It is generally available for leakage currents as low as 30 mA, which permits an earth-loop impedance of up to 1667 Ω before the shock voltage can reach 50 V, but is obtainable for 10 mA operation in high risk areas indoors such as schools or laboratories, or outdoors in gardens.

RCCBs are provided with test resistances operated by push buttons; they are usually made for load currents up to 100 A, which is ample for the load of most domestic installations. It is important to note however that earth fault loop testing will impose a heavy

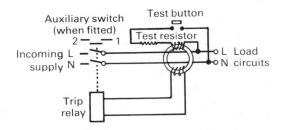

Fig. 63   Residual current circuit-breaker connections (*Crabtree*)

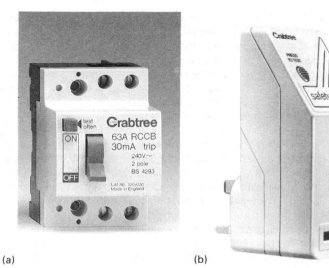

(a)   (b)

Fig. 64   Residual current circuit breakers, double pole
(a) 63 A; 30 mA; fixed type   (b) 13 A; 30 mA; plug in type
(Crabtree)

residual current on an RCCB which, at best, will cause it to trip thus
invalidating the test, or worse, will cause severe damage to the unit.
All RCCBs should be removed from circuit prior to carrying out
such a test.

In addition to fixed type RCCBs intended for installation in final
circuit distribution units, they can be incorporated into socket
outlets or even made as a portable and plug-in type. Typical fixed
and plug-in type RCCBs are shown in Fig. 64.

Earthing and equipotential bonding

The foregoing sections should make it clear that the primary means
of ensuring personal safety in the use of electricity and the preven-
tion of fire or damage due to electrical insulation faults is the
adequate and correct earthing and bonding of the electrical apparatus and adjacent conducting surfaces. Correct earthing ensures
that automatic protection by means of fuses or circuit breakers will
operate reliably and quickly and bonding ensures that no difference
in electrical potential or voltage can exist between adjacent conducting surfaces. In addition to these fundamental precautions,
other measures may be taken if necessary by means of a double

layer of insulation or physical separation by distance or insulating barriers between live circuits and earthed surfaces.

The efficient earthing of the metallic casing of electrical appliances and equipment is a primary safety precaution. This is to prevent an insulation fault causing a dangerous shock voltage to occur between exposed metalwork of electrical equipment and any other earthed surfaces (water taps, gas equipment, building structure, etc.) with which a person might be in contact at the same time.

Conduit, iron switch boxes, portable apparatus, cookers, water heaters, washing machines, motors, etc., should all have a proper earth connection. This is often provided by a three-pin plug and socket or a separate earth wire, which should be protected if liable to mechanical damage and must be run back to a good earthing point.

Fatalities have occurred to persons using unearthed or faulty apparatus in bathrooms, and special precautions should be taken to see that any exposed electrical metalwork and/or any other exposed metalwork liable to be in contact with electrical metalwork is properly earthed. The bath and water pipes may have to be bonded to earth in these circumstances. Socket outlets in bathrooms are unsafe and are not permitted, with the exception of special shaver supply units, the sockets of which are entirely isolated from the mains supply by means of a double-wound transformer, and in other rooms containing a shower cubicle no socket outlet may be within 2.5 m of the cubicle. Lampholders must be of the all-insulated shrouded type if within 2.5 m from a bath or shower cubicle, but totally enclosed light fittings are permitted. Any switches inside a bathroom must be out of reach of a person using the bath. Alternatively, these can be fixed outside the door, or ceiling switches with insulating cords can be used. Exposed elements of a fixed heater must also be out of reach of a person using a fixed bath or shower.

The wire size used for the main protective earthing conductors in house installations may be related to the size of the circuit cable with which it is associated, and must be insulated as for the circuit cable and coloured yellow-green. Assuming equal conductivity, it must not be less than the circuit cable size if this is smaller than 16 mm^2 section; not less than 16 mm^2 section if the circuit cable size is between 16 mm^2 and 35 mm^2; and half the circuit cable size for larger cables. In any case a separate earthing conductor must not be less than 2·5 mm^2 if mechanically protected or 4 mm^2 if not so protected.

Where an earthing conductor is taken to an earth electrode and is buried underground it should be protected against corrosion; if not protected against corrosion it must be not less than 25 mm^2 if of copper or 50 mm^2 if of steel, whether mechanically protected or not. If protected against corrosion but not mechanically protected it must be not less than 16 mm^2. (Aluminium is not permitted for this purpose.)

Non-armoured wiring generally contains an earthing conductor, which must be connected, together with other earthing wires, to an earthing terminal at all terminations. In larger sizes than 1 mm^2 the protective conductor may be smaller than it should be and may need supplementing to meet Regulation requirements. In plastic conduit systems, a protective conductor should be drawn into the conduits with the circuit wiring. Main water and gas pipes and central heating risers must be bonded to the main protective earthing conductor with cable of half the section of the main earthing conductor. However it should not be less than 6 mm^2 section and it need not exceed 25 mm^2 section copper except where protective multiple earthing (P.M.E.) of the supply is adopted and the consumer's main earthing terminal may be provided by the Supply Authority. This is called main equipotential bonding. Aluminium is not permitted for this bonding if condensation is likely to be present.

It is necessary to fix a warning notice reading SAFETY ELECTRICAL CONNECTION – DO NOT REMOVE at the main earthing terminal and where any other bonding conductors are connected elsewhere. Supplementary bonding conductors for earthing and bonding extraneous conductive parts in such places as bathrooms, etc., must be the same size as the associated circuit protective conductors with similar minimum sizes.

There are tables in the Regulations which also give the limiting impedance values for the circuit protective conductor corresponding to disconnection times of both 0·4 and 5 seconds for the circuit protective device.

7

Installation Accessories and Switching

In Chapter 6 we considered the mains and final circuits; now we have to consider the various fittings that may be employed at the ends of the circuits. In the house the majority of the fittings are for lighting, which may be from pendant luminaires or wall brackets. Decorative lighting is employed with some interior design schemes, but, as the considerations are purely aesthetic, it is not possible to give any guidance on these applications; but it is important to decide the positions of all the wiring points at an early stage. In laying out the installation for a new house, it is advisable to agree the position of the various lighting points *before* plastering and decoration. With prefabricated construction, definite channels are usually provided for all services, arrangements being made for looping-out to ceiling points for lighting and to socket outlets for power.

Ceiling-roses and plates

Modern outlet accessories are made of moulded plastics, and the ceiling-rose is designed so that it does not need a ceiling block. Two or more connectors have grub screws which grip the looped-in wires, and the ends of the flexible leads to the lamp go into separate terminals. To prevent straining the flex at terminal screws the separate ends of the flex must be passed through the bridge-piece or flex grip before it passes out of the ceiling-rose. The old-fashioned accessories with their small screws and washers, entailing overhead work, did not facilitate quick installation, so more convenient ceiling-roses are now made in which the flexible connections are made up before the ceiling-rose is fitted on site. A modern, white,

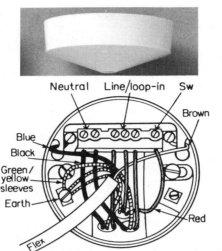

Fig. 65   White moulded ceiling-rose with looping and earthing terminals
(*M.K. Electric Ltd.*)

moulded ceiling-rose is shown in Fig. 65 in which are three circuit
terminals (two for looping-in as shown in Fig. 51).

With conduit work, ceiling-roses or metal ceiling plates are used
that can be mounted directly on to the lugs of a conduit box without
using a base block. A ceiling plate with hooks for the supporting
chains or tube suspension is employed with some heavy luminaires.
Bracket luminaires should have strong back-plates, and, when the
supporting tubes permit, the circuit cables should be brought right
in to the lamp-holder and sufficient lengths of wire should be left out
of the wall for this purpose.

Socket outlets and plugs

The British Standards Institution has laid down standard sizes for
sockets and plugs (British Standard Specifications 546 for round pin
and 1363 for flat pin types), and nonstandard sizes should not be
fitted. Nothing is more annoying than to find a variety of plug sizes
in a domestic installation, therefore all sockets for the same purpose
in an installation should be of the same size and type; they should
also be shuttered. The plugs must go well home into the socket so
that no live metal is exposed. Open, live, two-pin sockets are a
source of danger, especially to children. Three-pin types usually

have shutters actuated by the earthpin. Regulations prescribe shuttered sockets for domestic use and in particular recommend the
13 A flat-pin-type fused plugs and sockets to B.S. 1363 which it is
now standard practice to fit. One type of three-pin socket has an
inter-locking arrangement which holds the plug in, unless the switch
is in the 'off' position, so that the plug cannot be withdrawn when
the socket contacts are live. These, however, are not normally fitted
in domestic a.c. installations, usually being employed in industry or
for d.c. installations.

B.S. 1363 fused plugs are rated at 13 A, with fuses rated at 3 A
(red) and 13 A (brown) to B.S. 1362.

It is common practice for manufacturers to supply 13 A fuses
fitted into 13 A plugs, but it is important to replace these with 3 A
fuses where the plugs are connected to appliances of less than 700 W
rating.

The relative sizes of B.S. 546 round-pin plugs and sockets are
shown in Figs. 66 and 67. There are several makes of non-standard
sockets and plugs on the market – notably those made by Dorman
Smith Britmac Ltd. and Wandsworth Ltd. – which have different
arrangements and sizes of pins, but these are not normally used in
modern domestic installations. Wall sockets should be positioned
not less than 150 mm from the floor or above working surfaces to
avoid possible injury to the trailing flexible or impact damage to a
plug and socket when they are fixed vertically on walls.

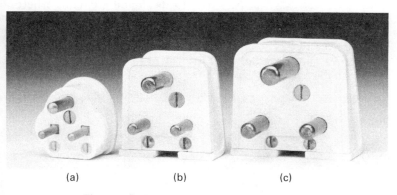

(a)   (b)   (c)

Fig. 66   Relative sizes of round pin plugs
    (a) 2 A standard
    (b) 5 A   ,,
    (c) 15 A   ,,

(*Crabtree*)

The 13 A socket is particularly intended for the a.c. radial and ring circuits described in Chapter 6 for house installations. It is made to fit on to round and square conduit boxes for surface or flush mounting, with or without a control switch or pilot lamp, and in multiple groups. For d.c. supplies each socket outlet must have a separate control switch because it is not safe to break the circuit by pulling the plug out, due to arcing at the contacts. A switch is not essential for a.c. supplies because arcing is negligible with zero current occurring twice in every cycle, although it is very convenient as it may save pulling out a plug to turn off an appliance. The wiring terminals are made to take 2·5 mm^2 and 4 mm^2 looped cables. The twin unit is increasingly useful, since it does not involve separate wiring for each socket and, with so many portable appliances being used in the home, avoids the use of the multiple adaptor. A well-designed installation should be provided with an adequate number of socket outlets, (bearing in mind that flexible leads fitted to such equipment do not normally exceed 1·5 m to 2 m in length), making such adaptors unnecessary, but many installations have such a limited number of socket outlets that the use of multi-outlet adaptors is often inevitable. It is, in fact, far better and safer to replace a single fixed socket outlet by a fixed multiple unit than to use an adaptor. Fig. 67 shows typical flush-type round pin socket outlets, and Fig. 68 shows flush-type 13 A switched socket outlets.

For portable lights and small-current appliances, sockets and plugs rated at 2 A can be wired to the lighting circuits where a convenient 13 A socket outlet is not available, but a fused connector is used for electric clocks, with an outlet hole for the flexible lead. It may be fixed by a central screw, since it is not intended for frequent withdrawal, and may be fused for 2 A or less; such a connector is shown in Fig. 69.

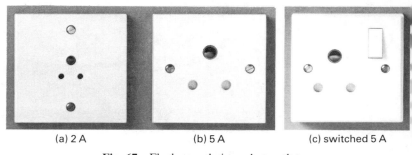

(a) 2 A   (b) 5 A   (c) switched 5 A

Fig. 67   Flush round pin socket outlets

(*M.K. Electric Ltd.*)

Fig. 68 Flush 13A switched socket outlets
(*M.K. Electric Ltd.*)
A plastic pattress is available for surface mounted outlets

If a socket outlet is intended for connecting equipment to be used outside the house, say for electric garden tools, a durable notice must be fixed on or near it reading 'FOR EQUIPMENT OUT-DOORS' and, in addition, the circuit or socket outlet must be protected by a residual current circuit-breaker (R.C.C.B.) having a rated operating current not greater than 30 mA, and an operating time not exceeding 0.04 s at a residual current of 150 mA.

The spur connection unit

This unit is used for taking a branch or 'spur' from a ring or radial circuit for a stationary appliance or 13 A socket outlets, and the

Fig. 69 Fused connector for electric clock
(*M.K. Electric Ltd.*)

Fig. 70   Switched fused spur unit with indicator lamp and flex outlet
(*M.K. Electric Ltd.*)

fuse, if fitted, must not be more than 13 A maximum nor exceed the rating of the spur cable. It can be switched and provided with an indicator light, and can also have an outlet for a flexible lead in the front plate. It is suitable for serving a water heater or a fixed fire, where a socket outlet is not appropriate. A flush, insulated, switched fused spur box with neon light and flex outlet is shown in Fig. 70. The number of fused spurs is not limited in standard circuits but must not be greater than the number of socket and other outlets directly connected in the circuit.

Lampholders

The lampholders employed for domestic lighting are of the bayonet type for tungsten filament lamps and should comply with B.S. 5042. Two side pins on the lamp cap go into slots, and the lamp contacts depress two spring-loaded plungers, with which they make contact. Small bayonet patterns (S.B.C. or B.15) are used for special purposes, including decorative schemes, but these should not be used for general purpose lighting. Standard bayonet holders (B.C. or B.22) are used for all lamps up to 150 W. For lamp sizes larger than 150 W, Edison screw holders are used (Fig. 71), and the lamp cap is provided with a screw thread, although 200 W lamps are also obtainable with bayonet caps. Although Edison screw lampholders are unusual in house installations, it is important to know that the centre contact must be connected to the line or phase wire of the circuit and the side or screw contact to the neutral wire. S.B.C.

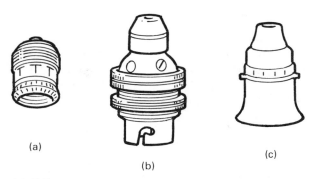

(a)

(b)

(c)

Fig. 71   (a) Edison screw lampholder
(b) Metal bayonet cap lampholder with shade ring
(c) White moulded B.C. lampholder with protective shield

(B.15) and S.E.S. (E.14) lampholders must not be used in circuits where the over-current protective device rating is greater than 6 A but B.C. (B.22) and E.S. (E27) lampholders and bi-pin types used for fluorescent lamps may be used in circuits with up to 16 A rated protection.

Brass lampholders that can be readily touched by a person standing on or in contact with earthed metal must be earthed; alternatively, insulated lampholders with protective shields that cover the lamp cap may be used. For domestic use 'all-insulated' lampholders made of moulded insulating material are commonly fitted in which there is an outer shell of bakelite covering the metal reinforcement of the lampholder. This type of lampholder lasts fairly well with low-powered lamps, though the heating effect of 100 W lamps and higher wattage may cause trouble with the shade ring, but heat-resisting types are made for temperatures greater than 165°C and are marked for a T2 temperature rating. These should be used with lamps in the 'cap-up' position in enclosed luminaires or those with doubtful ventilation. These lampholders are also shown in Fig. 71. In all lampholders provision should be made for gripping the flex, so that the lamp and shade are supported without straining the wires in the terminals. Bakelite lampholders and fittings should be periodically examined to see that the bakelite is not cracked or chipped. If arcing or flashing occurs over the surface of bakelite, it leaves it in a dangerous condition, as carbon tracks may be formed. These give conducting paths for the current, its insulating properties are impaired and it is no longer safe.

Shielded B.C. lampholders are fitted with a 'skirt' or shield of insulating material that covers the lamp cap and acts as the shade ring, which screws on to the bottom of the lampholder (as depicted in Fig. 71). Such shrouded lamp-holders are fitted in kitchens, lavatories, bathrooms and cellars where there is a possibility of personal contact with damp walls or ground when replacing a lamp.

Batten lampholders have a large circular flange or base for fixing direct to outlet boxes.

Adaptors

When it is required to connect more than one appliance to a single socket outlet, an adaptor plug is often employed, which may contain a fuse to limit the current taken from the socket outlets. The Regulations require that every portable appliance or light fitting must be fed from an adjacent or conveniently accessible socket outlet and do not recognise the use of adaptors, but adaptors will continue to be used in older and lesser equipped houses where there is a scarcity of socket outlets until future rewiring or extensions make them unnecessary. Safe designs of adaptors covered by British Standards are illustrated in Fig. 72.

Shaver supply units

The use of electric shavers adjacent to handbasins in bedrooms and in bathrooms, where shock risks are greatest and ordinary

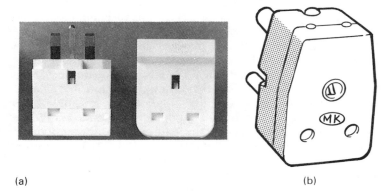

(a)   (b)

Fig. 72   (a) 13 A fused adaptor for B.S. 1363 sockets
       (b) 15 A and 5 A fused adaptor complying with B.S. 546

(M.K. Electric Ltd.)

socket outlets and portable appliances are not permitted, requires a specially safe socket outlet to be devised. Although electric shavers are generally of insulated construction, the shaver head is of steel and, since it makes direct contact with the skin of the face or neck, where shock could be painful and dangerous, an even safer method of protection than that by earthing is adopted by using a double-wound transformer to isolate the shaver from the mains. This also enables a choice of different voltages to be offered through separate sockets on the shaver supply unit, which is convenient for foreign visitors in hotels. The transformer core must be properly earthed, the shaver unit must incorporate a current-limiting device or a fuse not exceeding 3 A and the unit must comply with B.S. 3052. For use in places other than bathrooms, the unit must comply with B.S. 4573; the face plate should be marked FOR SHAVERS ONLY. A switched unit (which can be automatic) is desirable for units with transformers to avoid overheating of the windings when not in use. Shaver supply units are shown in Figs 73 and 74 as well as a 13 A adaptor for shavers for use in 13 A socket outlets.

Fig. 73   Shaver supply unit for bathrooms, flush type
(*M.K. Electric Ltd.*)

Switches and switching

A switch is a convenient piece of apparatus for opening or closing an electrical circuit. The single-pole switch makes or interrupts the supply on one pole or phase only; but when safety demands

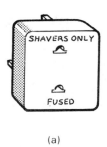

(a)   (b)

Fig. 74   (a)   Shaver adaptor, fused for 13 A socket outlets
(b)   Shaver supply unit for places other than bathrooms, flush
type

(M.K. Electric Ltd.)

complete disconnection from the supply, double-pole switches are used.

Gas-filled lamps take about one-fifth of a second to attain their operating temperature, during which time the current is decreasing from about six times its steady value, therefore with d.c. supplies a quick 'make' of the switch contacts is necessary so that heating and burning do not occur, and a quick 'break' is required so that the arc does not linger between the switch arm and the contacts; a long break is also desirable to assist in extinction of the arc. But on a.c. supplies the current is reduced to zero every half cycle, or 100 times per second so that special requirements to prevent arcing are not necessary, which is the reason why a.c. sockets and plugs may be used to disconnect without switches. A very simple switch mechanism is used in the a.c. switch by which the current is interrupted by silver contacts separating at slow speed, with a very small gap of 0·635 mm. This is called the microgap a.c. switch; the mechanism is silent in operation, without buffers or other damping devices, and has a long operating life. The action is quick 'make' and slow 'break', but these switches must only be used with alternating current. They are noiseless because the only moving parts are the operating dolly and the flexing contact blade. A section of this type of switch is illustrated in Fig. 75 (page 119).

Types of switches for lighting
Surface-type insulated switches which have an insulating cover over the live parts are commonly used. Sunk or flush switches with metal plates and dollies must have the metal parts earthed on to the

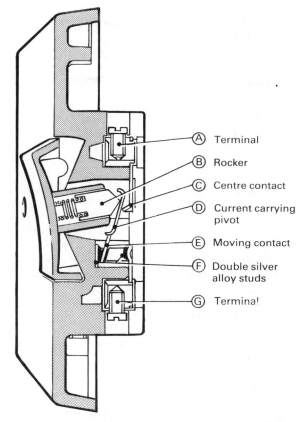

A Terminal

B Rocker

C Centre contact

D Current carrying pivot

E Moving contact

F Double silver alloy studs

G Terminal

Fig. 75 Section of patented mechanism of the Crabtree Rocker Switch
(*Crabtree*)

conduit box, and earthed switches must be used. The usual rating for domestic lighting circuits is 5 A, though they seldom have to carry this current since, on 240 V, a 100 W tungsten lamp takes only 0·42 A. Fluorescent lamps and their control gear form an inductive circuit and have an excessive starting current when switched on, the power factor is low, unless corrected to a higher value with capacitors, and consequently, when determining the control switch capacity twice the normal load current must be allowed.

Although there still are many tumbler type switches for a.c. to be found in the older houses, the rocker type a.c. switch is now almost

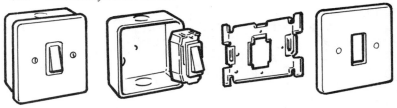

Assembly = Box + Switch + Yoke + Plate

Fig. 76   Surface type rocker switch assembly

(*Crabtree*)

universal and very popular due to its smooth and easy operation. Instead of having a snap-action operating dolly, a rocking-action lever works the switch by simply pressing in the projecting end of the rocker, which has only a small projection through the switch plate. This type of switch occupies such a small space that groups of two and three switches can be accommodated in a single square switch plate and box. They are narrow enough to use in the hollow spaces of the metal door frames and are also used in 13 A switched socket assemblies. This type of switch is illustrated in Figs 76 and 77. Interesting features of this design are cable terminals in which the wires push in and are held securely without screws (but with release buttons) and a clip-on cover plate surround which can be easily removed for room redecoration (see Fig. 77).

Suspension switches are suspended by a flex from the ceiling. This type of switch should not be fitted when it can be avoided, since the

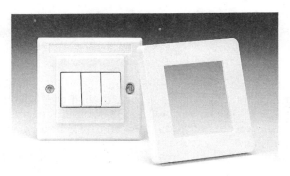

Fig. 77   Group of three rocker switches, flush type, and right, clip-on cover plate surround

(*Crabtree*)

Fig. 78 Ceiling switch

supply voltage is on the conductors of the flexible cord even when the switch is off and trouble is often experienced when the insulation of the flex becomes frayed and worn.

This type of switch is also incorporated in switched lamp-holders, and the same criticism applies if the lighting point is not switched elsewhere in the room. But where a switched lampholder on a lamp-standard is connected to a socket outlet, this is accepted as a convenient arrangement, although the flex should be inspected frequently for damage and wear.

Ceiling switches are preferable to and safer than suspension switches previously mentioned. The ceiling switch is operated by a pendant cord, which is pulled either to switch 'on' or to switch 'off' consecutively. Such a switch is illustrated in Fig. 78. This type of switch is frequently used in bathrooms, where all switches must be out of reach of a person in the bath or shower and is useful in bedrooms so that the light can be controlled from the bed. Compared with a wall switch, a considerable amount of wiring can be saved, as the switch wires down the walls are dispensed with. This is shown in Fig. 79 in which the circuits for wall switches and ceiling switches are compared.

Push-button switches with consecutive action are often fitted to table lamps and suspension switches, and have a single projecting button which is pressed for either 'on' or 'off' as required. Push-button switches with reverse action are sometimes fitted to deep cupboard doors so that, when the door is opened, an internal light is switched on; closing the door switches the light off.

Switching circuits

One-way, single-pole switching has been illustrated when looping-in was considered, and various other switching arrangements will now be discussed.

The circuit diagram illustrated in Fig. 80 (page 123) shows a one-way switch with an additional 'loop-in' terminal for use where the cable run passes through the switch position. This avoids fitting a connector in the outlet box, which is not good practice; to facilitate the use of twin cable it is much better to use a three-terminal ceiling

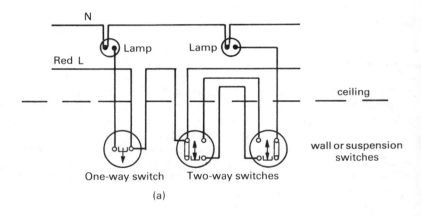

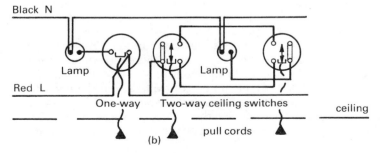

Fig. 79 Economy in wiring with ceiling switches (no switch drops in walls)
   (a) Typical wiring behind ceiling with drops to wall switches
   (b) All wiring behind ceiling

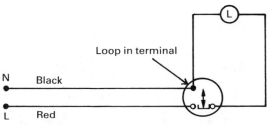

Fig. 80 One-way switch with 'loop-in' terminal

rose for the purpose. This type of switch is of interest because it provides a looping-out point for another point or an extension later.

All switches with metal parts that can be touched or are in contact with metal switch plates must be earthed, and outlet boxes or enclosures must be provided with an earthing terminal for this purpose, even if an insulated switch is fitted in the first instance. Where metal switch plates are employed, this feature should also earth the plate or an earthing yellow/green 'pigtail' must be fitted.

Fig. 81 shows three lamps, of which one, two or three can be in use at any time by using two switches.

The double-pole switch is shown in the circuit of Fig. 82. Where neither pole of the electricity supply is earthed, linked double-pole switches must be used throughout a two-wire installation and in other installations for all heating appliances in which heating elements can be touched and which are not connected by means of a plug and socket. Thereby the danger of a single-pole switch in the black wire cannot occur, as the apparatus is entirely disconnected from the supply when the switch is 'off'. Double-pole switches can also be used for simultaneous control of two separate circuits.

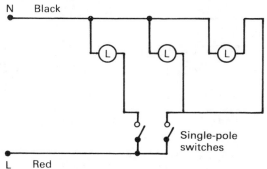

Fig. 81 Wiring to control a three-light luminaire

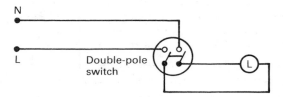

Fig. 82   Connections for a double-pole switch

TWO-WAY SWITCHING   This is a convenience that is very useful in bedrooms, on the staircase, in corridors and in rooms with two doors. A two-way switch is a single-pole changeover switch, and it is possible to waste a considerable amount of cable by incorrect connection. The proper method of wiring is illustrated in Fig. 83 at (a), and (b) shows the wrong method of inter-connecting two-way switches in which both live and neutral leads are brought into the switches; the risk of a short-circuit is ever present, and if the switch

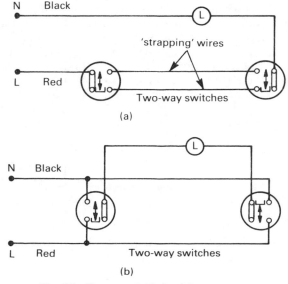

Fig. 83   Two-way switch wiring
(a) Correct method
(b) Incorrect method

cover is removed by anyone it is dangerous. Regulations prohibit such an arrangement by not permitting a single-pole switch in the neutral line where this pole of the supply is earthed, but this method is illustrated in order to emphasise its danger.

INTERMEDIATE SWITCHING   Long halls, corridors and passageways with many doors call for switching the same group of lights off and on from more than two positions. This is done by using intermediate switches. A circuit embodying one intermediate switch is shown in Fig. 84, but any number of intermediate switches can be connected in the same circuit by interposing them in the two 'strapping' wires between the two-way switches and connecting them in exactly the same way as the first intermediate switch shown.

MASTER CONTROL SWITCHING   This does not require a special type of switch and standard switches are employed. The master switch is so connected to the circuit as to have overriding control over the subsidiary switches. Fig. 85 shows a lighting circuit with each light individually controlled by a single-pole switch but with a single-pole switch in master control of the circuit as a whole. A double-pole linked switch could be used as a master switch so as to isolate both red and black leads of the circuit if the need was considered great enough.

There are many other switching variations for special purposes, such as dimming control and restrictive lighting, but any good installation contractor can devise special circuits with suitable switches for special requirements.

Dimmers and time switches

A dimmer switch for room lighting is now a popular facility in the home. It has great advantages in dimming room lighting to a

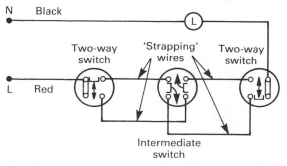

Fig. 84   Circuit with one intermediate switch

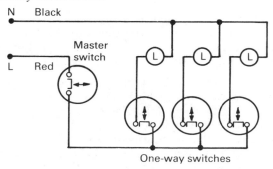

Fig. 85   Master control of three lamps

suitable low intensity for television viewing and for night lighting in nurseries or hospital wards. These units use electronic circuitry to reduce the voltage of the circuit, which is extremely effective and economical for lighting purposes, and some makes of dimmer switch can be accommodated in existing switch boxes to take the place of a switch. For television viewing, dimming is a great advantage because it avoids making the TV screen too bright in order to compete with normal room lighting, which is not good for the eyes or the TV set.

Time switches are being increasingly used in domestic a.c. installations nowadays. This type of control consists of an electric synchronous clock, which operates contacts 'on' and 'off' at predetermined times as set on an indicator dial. The most important use is for timing off-peak heating installations; and other minor applications are for time control of electric blankets and tea-makers for which plug-in-type units are available for use in socket outlets without any special wiring connections. Time switches are also incorporated in electric cookers for oven control.

In all cases of switch control, whether manual for lighting, heating or power, or automatic for time switching, it is always important to determine and specify the current rating required in each case, because this can range from the small 5 A lighting switch to the large 60 A or 100 A capacity control for a heating installation and too small a switch will overheat, burn at the contacts and break down very soon. The largest capacity controls usually have a high-capacity, solenoid-operated contactor switch controlled by a low-rated time switch.

Electronics has entered the field of domestic controls and while these systems have been used in luxury premises for automatic and remote variable control of lighting, doors, curtains, alarms and other services, ingenious and cheaper electronic devices are becoming available and will be used to a greater extent in more houses in the future. A new device called a Triac enables almost any form of control to be devised together with appropriate sensors; it is so small that it will go into a wiring outlet box and can be operated by only 9 V with very light insulated wiring. This will certainly revolutionise house wiring of the future but the methods of application will need working out and to be regulated for safety reasons before they become common practice. Electronic touch buttons, now used for calling lifts in tall buildings, are likely to be developed for lighting controls with no moving parts and will no doubt become popular for house installations. In these, a sensitive front plate the size of a postage stamp acts as a capacitor when a finger approaches it closely and triggers an electronic circuit which switches on the lights or any other equipment. The control of electrical and electronic circuits from remote positions by means of infra-red or ultra-sonic signals transmitted from hand held touch-button units will soon become commonplace as will digital electronic signals superimposed on the normal mains wiring around the house to control units plugged into socket outlets. All these systems may be developed further and used to control the 'All Electric/Electronic House' of the future under the command of a central domestic computer.

8

Lighting

Early man used torches lit from the camp fire for illumination and this primitive method of indoor lighting was succeeded by the burning of resinous wood and combustible oils and fats in various containers. But these early forms of artificial lighting could not match daylight visibility and it was not until the nineteenth century that more ingenious lamps for getting better light from oil were developed, and by the end of the century the use of gas as a lighting agent became widespread and the electric lamp had become a viable proposition. As a result of the Industrial Revolution and the setting up through legislation of minimum standards of lighting for factories and work premises, electricity began to be used as a commercial source of lighting towards the end of the nineteenth century. With the rapid development of electric lighting the Lighting Industry Federation, the Electricity Council and the Chartered Institution of Building Services Engineers have published much information on the proper use of electric light for different purposes and legislation has set up minimum standards of lighting for factories. The latter body continues to publish recommendations for lighting building interiors with methods of calculation and design and tables of illuminance for various kinds of work, in a Code of Practice revised from time to time as trends towards better lighting and higher lamp efficiency raise the standards of lighting. The values given in this chapter are taken from the latest edition of this Code.

The maximum illumination intensity for seeing is, of course, that in which the eyes can see best in comfort; but since the eyes have been attuned by nature to the brightest daylight conditions, the equivalent of this in artificial lighting cannot be economically obtained without creating excessive glare which is detrimental to eyesight. So modern illumination aims at comfortable visibility with a light intensity on the object viewed that enables the smallest detail to be seen with ease and with daylight intensity as the ideal optimum value.

Electric lamps

The earliest electric lamp for use in houses consisted of a carbon filament in an evacuated bulb. This was followed by the more efficient metal filament vacuum lamp and later by the gas-filled lamp. Modern general service lamps have tungsten filaments, but the bulb contains an inert gas consisting of nitrogen and argon. This enables the filament to be run at a higher temperature and gives greatly increased efficiency compared with the earlier lamps for 40 W and above. Even so, the light energy is a small fraction, only about 6%, of the total energy, as shown in Fig. 86. With gas-filled lamps the filament is in the form of a fine coil of wire less than a thousandth of an inch in diameter. The coil is formed in a close spiral to minimise the cooling effect of the gas. A later filament development was in the 'coiled-coil' lamp, which is even more efficient. This is due to a greater concentration of incandescent metal, which emits more light for the same current as a single-coil

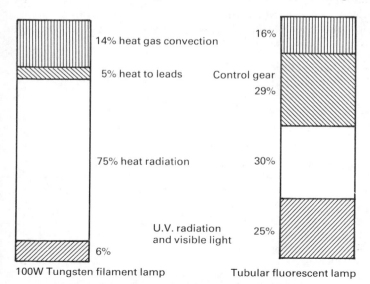

Fig. 86 Energy distribution in domestic lamps

Fig. 87   Coiled-coil filament

lamp. The filament arrangement is illustrated in Fig. 87. Table 10 shows the lumens output of standard tungsten filament lamps which are the average values throughout their useful life.

Light units

Lumen
A lumen (lm) is the unit of light flux used in describing the total quantity of light emitted by a light source or received by a surface.

Lux
This is the metric unit of illuminance value and is equal to one lumen distributed over one square metre (0·093 lumen per square foot).

Candle-power – candela
The candle-power of a source is an old-fashioned term designating luminous intensity, which is now expressed in candelas (cd); a light source of 1 candela emits 4π lumens (12·57 lm).

Mean spherical illuminance
This is expressed in lux but with a somewhat different meaning to that given above. It is the average illumination over the internal surface of a small sphere centred at the point of the light source, i.e. it is the incident flux on the surface of the sphere divided by the area of the sphere.

Table 10   Lumens output of tungsten-filament lamps at 240 V

Type of lamp	Lamp watts					
	25	40	60	100	150	200
Single-coil: lumens	200	325	575	1160	1960	2720
Coiled-coil: lumens	—	390	665	1260	2040	—
Light increase %:	—	20	16	11	4	—

Note: 110 V lamps have approximately a third greater output than 240 V lamps in 40 W and 60 W sizes (single coil).

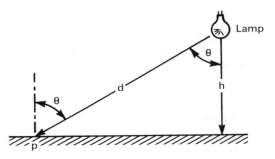

Fig. 88   Illuminance from a lamp

Illuminance at a point

The illuminance received at any point is inversely proportional to the square of the distance of the light source (Inverse Square Law) and proportional to the cosine of the angle of incidence. This is illustrated in Fig. 88, in which a lamp with a light intensity of I candelas in the direction shown is suspended h metres above the horizontal plane and d metres from the point P to the lamp. The illuminance (E) at point P is given by:

$$E = \frac{I \times \text{Cos } \theta}{d^2}\text{lumens/m}^2\text{, or lux.}$$

As $\text{Cos } \theta = \dfrac{h}{d}$, $\therefore E = I \times \dfrac{h}{d^3}$ lux.

The lux or lumens per square metre is the unit of illuminance generally used in room lighting calculations and is employed for measuring the lighting obtained. Common values of illumination

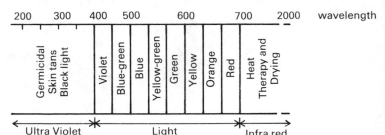

Fig. 89   Colour Divisions of the Visible Section in the electromagnetic spectrum

range from 0·0027 lux for 'starlight' and 0·2 lux for full moonlight to such values for interior lighting in domestic premises as 50–300 lux. Artificial lighting has far to go, considering that daylight is of the order of 10,000 lux. Fig. 89 shows the colour constituents of visible radiation in the electromagnetic spectrum and the useful radiation in the adjacent areas against a scale of wavelengths.

Light measurement

The human eye cannot measure illuminance, though it can compare equality of brightness with fair accuracy in photometers, in which the light is compared with a standard source.

Portable photometers or lightmeters are now extensively used to

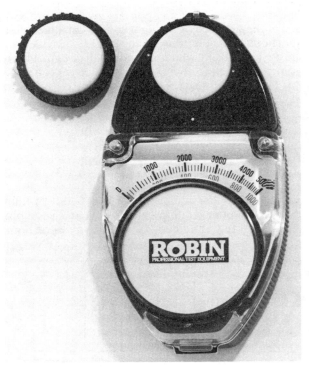

Fig. 90   Lightmeter, which gives illuminance in lux, with multiplier attachments for the higher scales

(*Robin Electronics Ltd.*)

give direct readings in lux. Their action depends upon the effect of the light on a photoelectric cell; this generates a very small current which gives a reading on a milliammeter calibrated in lux. Such a lightmeter is illustrated in Fig. 90, and by its use measurements can be taken to check any lighting system when installed or at a later date to show the effect of dirt and ageing of the lamps. Before the introduction of international standards, lightmeters were calibrated in foot-candles. The foot-candle was the unit of illumination, which was 1 lumen distributed over a surface area of 1 square foot, and it is still used in the USA.

Types of incandescent lamps

The bulbs are made of clear glass in all sizes, and with the internal surface frosted (pearl) up to 150 W. The clear lamp should not be used where there is possibility of glare, which will cause eyestrain. Clear glass lamps should be enclosed in suitable shades or diffusing enclosures. Pearl lamps give the same light output as clear lamps, but the diffusion obtained from the frosted glass lessens the glare and softens the shadows, particularly at low mounting heights. The choice of clear or frosted glass lamp bulbs depends on the kind of shade used, whether diffused (for general lighting) or sharp shadowing of objects or fine work being done is desired; the lamp fitting may need a concentrated light source to make prismatic glass sparkle, for example. Opal lamps have a thin skin of opal glass on the outside of the clear-glass bulb, but this type of lamp has been largely superseded by the white, internally-coated bulb. The diffusion is complete and shadows are minimised but the light output is reduced by about 10%.

Daylight-blue lamps have the bulb made of a blue-tinted glass. The glass absorbs some 50% of the total light, so about double the wattage is often needed. Such lamps are better for colour matching, but the light is sometimes considered cold and depressing. Coloured lamps are available with bulbs internally coated blue, green, white, amber, red, pink or yellow. These all absorb high proportions of the light output – according to colour – of the filament, which makes them very inefficient for lighting purposes. The mushroom-shaped lamp has attained much popularity due to its attractive shape and smaller dimensions than the general-purpose lamp. It is internally coated white or other tint and provides an evenly diffused light.

A number of decorative types of lamp are also available in various shapes: small, round bulb; plain and twisted candle-flame shape; and long, tubular candle shape. These are suitable for some styles of decorative luminaires, but the lamps are of the inefficient, low-wattage, vacuum type. So-called 'long-life' lamps are simply under-run lamps, i.e. designed for a slightly higher voltage than the standard lamp rating with reduced light output and correspondingly longer life (see p. 36).

The incandescent filament lamp is relatively inefficient and gives from 8 to 14 lumens per watt, depending upon its size, for the normal domestic range of lamp wattages; and it has a rated life of 1000 hours but it may last much longer with decreasing light output. But it is cheap, easy to install and can be conveniently controlled by simple switches.

The latest development in metal filament lamps is the tungsten-halogen lamp in which the presence of iodine prevents the volatil-ised filament from being deposited on the inside of the bulb and causes it to be returned to the filament, so preventing blackening of the bulb and increasing the life of the lamp and its light output. These lamps are occasionally used for domestic lighting but they are commonly used as high wattage lamps of small dimensions for photographic projectors and equipment and for floodlighting.

Electric-discharge lamps

An electric discharge through gases gives a much greater light efficiency than can be obtained with incandescent lamps. Discharge lamps are best suited and usually designed for operation on a.c. supplies, but with suitable control gear they can also be used with d.c.

Mercury (bluish) and sodium (chrome-yellow) electric-discharge lamps are used for street lighting and factory installations, but the single colour is a disadvantage, making them unsuitable for domes-tic and commercial uses. To overcome this the mercury discharge lamp bulb has been coated internally with phosphor, which converts ultra-violet radiation produced by the discharge into visible red light, so improving the colour rendering and producing a cold, bluish-white light which has been much preferred for street and factory lighting. These lamps are mostly produced in the larger sizes for this purpose, and although 50 W and 80 W lamps are obtainable the light is not considered 'warm' enough for home use. The efficiency is about four times higher than that of the gas-filled tungsten lamp and the rated life is 7500 hours.

The high-pressure sodium lamp which is made to give a warm golden light has become most popular for street lighting due to its higher efficiency and longer life.

The tubular fluorescent lamp has been developed in a range of warm and cool colours, wattages and lengths, and, though more expensive initially than the tungsten lamp, the much higher light output obtained proves economical if the lamp is in use for long periods. The average light output for the first 5000 hours is 60 1m/W for 'warm white' and 57 lm/W for 'daylight' colours of a 1500 mm (5 ft) lamp rated at 80 W. The corresponding figures for a 1200 mm (4 ft), 40 W tube are 65 and 62 lm/W. Some colour variations are obtained by a coating of different phosphors and filter coats on the inside of the tube, but this reduces the light output.

The lumens obtained from 240 V fluorescent lamps, ranging from 15 W to 85 W in warm white and natural white type (these lamps being commonly used in domestic premises) are given in Table 11. The complete range lists fifteen sizes from 4 W to 125 W.

The lamp makers have reduced the long list of fluorescent lamps previously made to only five new standard lengths. These have 26 mm diameter tubes, Krypton filled, with superior phosphors and higher efficiencies, but alternative tubes (38 mm) for replacements of the older lamps are available where the new standard lamps are unsuitable. B.C. lamps, reflector lamps, lamps with an earth strip, 40 W 600 mm and 85 W 2400 mm lamps will be obsolete. Three of the sizes that will be mostly used in the house are:

Length mm	Watts New	Replacement
1500	58	65
1200	36	40
600	18	20

Table 11   Lumens output of fluorescent lamps

Fluorescent Lamps			Approximate Light output, lumens at 2000 hrs. operation	
Lamp Watts	Nominal length mm.	Load with control gear	Warm white	Natural
15	450	25	800	600
20	600	32	1100	800
30	900	40	2150	1600
40	1200	48	2700	2100
50	1500	70	3550	2400
65	1500	78	4600	3400
75	1800	91	5650	4000
85	2400	100	6750	5000

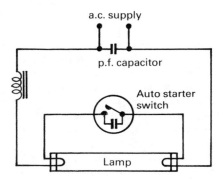

Fig. 91   Fluorescent lamp with glow starting switch
(All the starting equipment is normally contained in the base unit of the
luminaire)

The light output is about 11% greater for the same wattage of the older lamps.

Some auxiliary apparatus is required, as shown in the circuit diagram of Fig. 91; this may consume an extra 15 watts or so which should be taken into account when considering loads or the efficiency of the lamps. The automatic starting switch may be of the 'glow' or 'thermal' type; the small condenser across it is to suppress radio interference. The starting device heats the lamp electrode filaments, and then opens and initiates the discharge at a high voltage from the choke, but the final running voltage across the lamp is about 110 V. The choke, on a.c. supplies, limits the current to the correct value but lowers the power factor, so a mains capacitor is fitted in parallel across the supply to improve the power factor. The 80 W lamp requires about 7·5 µF (microfarads) and the 40 W lamp about 3·2 µF capacitors. These starter circuits take several seconds to strike the arc. 'Instant start' lamps are available which have different control gear and circuits, and an 'earthing strip' along the length of the lamp which is necessary for reliable starting. Failure of the lamp is usually due to deterioration of the fluorescent powders in the lamp coating and evaporation of the electrodes; starter switches can be replaced, but failure is accelerated by frequent switching, and insufficient preheating prevents the lamp from starting.

The electric discharge between the electrodes in the tube goes out 100 times a second on a 50-cycle supply. This causes a stroboscopic effect when the light is reflected on rotating or moving objects at critical speeds. To overcome this a twin lamp circuit can be used on a single-phase supply; the circuit is shown in Fig. 92. The upper lamp goes out at a different time in the cycle from the lower lamp, due to the different circuit constants, which puts their operation out of phase with each other, and the overall power factor is corrected automatically to nearly unity by this circuit.

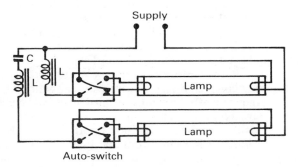

Fig. 92   Twin lamp circuits; eliminates stroboscopic flicker
(C = condenser, L = choke)

Philips Electrical Ltd. originally produced a fluorescent luminaire with a tungsten lamp acting as the choke for the fluorescent lamp. This saved cost and weight in the fitting, and the combination produces its own warmer colour rendering with slightly lower efficiency. A mercury discharge lamp is now made with a self-contained tungsten filament in the bulb which improves the colour rendering and makes additional control gear unnecessary.

Fluorescent tubes are also made in circular form, 406 mm (60 W & 40 W) 305 mm (32 W) and 209 mm (22 W) diameter, which are convenient for conventionally shaped luminaire designs. The latest development in discharge lamps is a 16 W 12 mm diameter fluorescent tube with integral starter, formed into a flat compact shape about 140 mm square which gives a light output almost equivalent to a 100 W incandescent lamp but with five times the life. Other types and shapes of tungsten, fluorescent and new lamps are constantly being developed.

Domestic lighting

There are three main requirements in home lighting: sufficient uniform illumination, freedom from glare due to unscreened lights or reflection from polished surfaces, and freedom from deep shadows and abrupt contrasts. The light should not flicker or vary in intensity, but this is not apparent on 50-cycle systems with metal-filament lamps and is usually not noticeable with fluorescent lamps especially if the twin lamp circuit described earlier is used. The choice of fittings and shades depends on personal tastes and what kind of lighting is desired, but it is most important to ensure that luminaires have adequate ventilation and not to fit higher wattage lamps than they are designed for, otherwise serious fire risks will arise from over-heating, particularly with filament lamps.

The colour and surface of the walls have direct bearing on the effect obtained. With direct lighting most of the light is directed downward by suitable reflectors, while with indirect lighting the source of light is concealed and illumination is obtained by reflection from ceilings, walls, curtains or other surfaces with poor efficiency. Indirect lighting is not used to any great extent in houses, except for decorative effects in addition to other methods of direct lighting. Semi-indirect lighting is obtained with translucent bowls and other types of luminaire with opaque bottom panels and translucent side panels which considerably reduce the glare associated with direct lighting, especially when the shade employed fails to cut-off direct light from the lamp.

With semi-indirect lighting, such as that with a thick diffusing bowl, or direct lighting with large silk shades or other inefficient types of luminaire, an ample number of socket outlets should be provided for desk or table lamps and lamp standards. In sitting rooms where reading is done, at the dinner table and at dressing tables not less than 100–150 lux should be provided. In kitchens, wash-houses and bathrooms totally enclosed diffusing luminaires should be used so that the lamp is protected against steam and dirt, and the exterior of the globes can be kept clean. Staircases and passages should be sufficiently illuminated in order that there is not too much contrast in coming out of a well-lighted room. An ample number of socket outlets should be put in at the time of installation, as they are not only essential for the multiplicity of portable electric equipment found in most homes nowadays, but they are often needed for the variety and number of luminaires that are frequently included in ultra-modern lighting schemes to obtain variable effects. Multiple plug adaptors with long trailing flexible leads are unsightly, and extra socket outlets suitably spaced around a room provide a more convenient arrangement and a safer installation. 50 lux is recommended for general illumination on the working plane in living rooms and at floor level in bedrooms and garages, with 150 lux locally for casual reading, at bedheads and for halls and landings. Local illumination at desks in studies, for sewing or machining and on kitchen and work-shop benches should be 300 lux. In bathrooms and on stairs 100 lux is adequate.

Colour

Artificial lighting contains different proportions of spectrum colours from daylight. Metal-filament lamps contain a larger proportion of red and yellow rays, and are deficient in green and blue. Colours in which red and yellow predominate appear much warmer under artificial light than in daylight, while green and blue surfaces appear dull because they absorb much of the light. The idea that green is a restful colour is only true when ample illumination is provided, since eyestrain may result if too much light is absorbed. With dark colours considerably more light is necessary for good visibility than for light colours, whether referring to objects being inspected or to the surrounding walls and ceiling of a room which contribute their quota of light by reflection.

The ratio of reflection under a tungsten lamp to that with daylight, for different-coloured papers, is given in Table 12 below.

The amount of reflection also depends on the surface texture. A glossy enamel finish will reflect more light than a flat tone or matt surface. Glossy surfaces give specular or directional reflection and matt surfaces give diffuse reflection; dark, matt surfaces absorb most of the light.

Good mirrors and best white surfaces reflect about 90% of the incident light, and some approximate reflection factors are given in Table 13.

Table 12 Colour reflection ratios of tungsten lamp to daylight

Colour of paper		White dif-fusing	Red	Orange	Yellow	Green	Deep blue
Ratio:	$\dfrac{\text{Reflection under tungsten lamp}}{\text{Reflection with daylight}}$ =	1·0	1·48	1·26	1·08	0·75	0·69

Shadow

The perception of an object in three dimensions is affected by the light and shade effect, as well as colour difference. There should be

Table 13   Reflection factors

Material (clean)	Approx. reflection factor	Material (clean)	Approx. reflection factor
	%		%
White tile, glossy	80	Ivory, matt	64
White paint,		Light stone	58
glossy	78	Middle stone	37
Plaster, matt		Yellow brick	35
white	70	Red oak	32
Ivory, glossy	69	Red brick	25

sufficient contrast of brightness between the object and its surroundings for easy vision, and the shadows must not be dense and black but soft and grey. Heavy shadows are dangerous as they are liable to cause accidents, especially on stairs and in passages. Direct lighting gives more pronounced shadows, while with indirect lighting there is an absence of hard shadows. Semi-indirect lighting combines both effects, with soft shadows from a bright ceiling and a degree of harder shadows from the light fitting.

Illumination calculations

To calculate approximately the size of lamp necessary to provide the required luminance in a room, the following formula can be used:

Lumens per lamp
$$= \frac{\text{Average light flux required (lx)} \times \text{Area per lamp (m}^2)}{\text{Utilisation factor} \times \text{Maintenance factor}}.$$

This is the simple 'lumen' method, and assumes correct location and spacing of lamps, each one in the centre of its own area; but more advanced methods are used if necessary for industrial and commercial schemes. Where the working surface is a wall or other vertical surface a different computation is used in which the average cylindrical illumination of each fitting and other reflection factors are involved, but this aspect need not concern the householder.

Utilisation factor
This is the ratio of the total flux received on the working plane to the total flux produced by all the lamps in the room. Sometimes called

the coefficient of utilisation, its value varies from about 0·3 to 0·5 with direct lighting in average conditions and rooms in houses and depends on the type of luminaire used, the dimensions of the room, and the colour and condition of the walls and ceiling.

The utilisation factor is obtained from tables giving the classifi-

Table 14 Utilisation factors

Light fitting	Ceiling	Fairly light 50%		Very light 70%	
(Tu = tungsten lamp) (Fl = fluorescent lamp)	Walls	Fairly dark 30%	Light 50%	Fairly dark 30%	Light 50%
	Room index	Utilisation factor			
(a) Open opaque reflectors 75% light downwards (Fl or Tu)	0·6	0·31	0·35	0·31	0·36
	0·8	0·4	0·44	0·4	0·45
	1·0	0·44	0·49	0·45	0·49
	1·25	0·49	0·53	0·49	0·55
	1·5	0·53	0·57	0·54	0·58
(b) *Bare lamp on ceiling batten; 65% light downwards (Fl)	0·6	0·22	0·27	0·24	0·29
	0·8	0·3	0·35	0·31	0·37
	1·0	0·35	0·4	0·37	0·44
	1·25	0·4	0·45	0·42	0·49
	1·5	0·44	0·5	0·47	0·54
(c) Diffuser with open top and louvred beneath; 30% light downwards (Tu)	0·6	0·2	0·24	0·23	0·28
	0·8	0·26	0·3	0·3	0·35
	1·0	0·3	0·34	0·34	0·4
	1·25	0·33	0·38	0·39	0·45
	1·5	0·36	0·41	0·44	0·49
(d) Enclosed diffusers spherical or near spherical; 45% light downwards (Tu)	0·6	0·16	0·2	0·18	0·23
	0·8	0·22	0·27	0·24	0·3
	1·0	0·26	0·31	0·29	0·36
	1·25	0·3	0·35	0·34	0·41
	1·5	0·34	0·39	0·39	0·45
(e) Complete luminous ceiling of translucent corrugated strip or pan-shaped panels (Fl)	0·6	0·15	0·2	(ceiling cavity should be white and cavity depth not more than a third of the width)	
	0·8	0·25	0·28		
	1·0	0·31	0·34		
	1·25	0·34	0·37		
	1·5	0·36	0·4		

* The addition of a diffuser reduces the downward light to about 50% and reduces the utilisation factors by about 10%.

cation of various types of luminaire, according to their directional light distribution in the BZ (British Zonal) system of classification, which is given by reputable manufacturers for each fitting they make, and by applying factors depending on room size (room index) and reflectances. This can be studied in more detailed works on illumination, particularly the technical reports of the Chartered Institution of Building Services Engineers, but reasonably satisfactory illumination calculations can be done by the simpler method of using the tabulated values given in Table 14, for typical fittings shown in Fig. 93, which are based on a much earlier and less complex form of assessment for lighting schemes but none-the-less still reasonable and more easily understood and workable by the reader of this book. They are a simplification of the C.I.B.S.E. table of utilisation factors which are now more detailed and based on specific numerical values of wall and ceiling reflectances.

Room index
The room index is first found by working out the expression:

$$\frac{\text{Room length} \times \text{Room width}}{\text{Height above working plane} \times (\text{length} + \text{width})}.$$

For example, if the height is 2 m, and the room is 6 m long and 5 m wide, the room index is:

$$\frac{6 \times 5}{2(6 + 5)} = 1 \cdot 36.$$

The dimensions of a room affect the utilisation factor, because in a small room a large proportion of the light is absorbed by the walls, whereas in a large room with a number of lights a much larger proportion of the light from the lamps falls directly on the working plane.

Effect of walls and ceiling
The colour and surface of the walls have a considerable effect on the illumination on account of their reflecting power. The lighter the colour of the walls and ceiling, the more light is reflected and, consequently, the higher the utilisation factor. Allowing a small proportion for deterioration, wallpaper colours can be arranged as below in three classes: light with 50% reflection or more, medium with about 30% reflection and dark with about 10% reflection. The colours are arranged in order of reflecting power, and this should be taken into consideration when deciding the total light required.

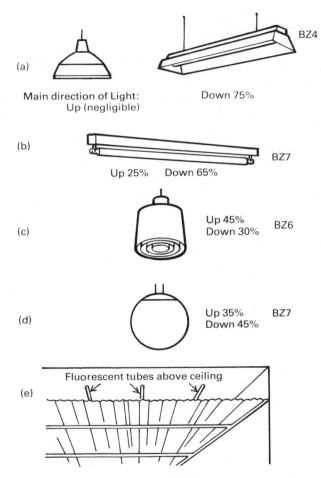

Fig. 93 Typical luminaires and BZ classifications

(a) Open opaque reflectors. High efficiency
(b) Bare fluorescent lamp on ceiling batten base. Diffuser added reduces glare
(c) Open top diffuser with louvres beneath. Provides glare-free illumination
(d) Enclosed diffusing luminaire. Softens shadows
(e) Luminous ceiling with translucent strips or panels. Very even shadow-free lighting

Light: white, cream, yellow, light orange, light stone, light buff and pale tints.

Medium: grey, pink, light green, sky blue, stone.

Dark: dark grey, brown, red, dark green, blue.

Ceilings can be divided into very light – generally white or cream and very pale tints – with 70% reflection; fairly light with 50% reflection (as for light walls in Table 14) and dark with 30% reflection (as for fairly dark in Table 14) but a column for this is not shown because domestic room ceilings are not usually dark. The effect of smoke and dirt will lessen these figures.

Maintenance factor

This allows for the falling off of efficiency due to deterioration of reflectors, walls and ceilings caused by dirt. With cleaning every few weeks, the average illumination obtained will be about 80% that of the original clean and new conditions, and a maintenance factor of 0·8 is usually taken for normally good conditions. This can be increased for extra-clean conditions but can fall to as low as 0·4 or 0·5 in dirty industrial surroundings.

Table 14 gives typical basic forms of luminaires illustrated in Fig. 93, together with the related utilisation factors for various room index values. These five sections and figures should meet most domestic applications.

The typical luminaires illustrated are simple diagrams, but some more decorative domestic lamp shades may give results that differ considerably from calculated values.

To ascertain lamp size

From the foregoing the lamp size for any room can be worked out, and an example will now show the application of the method described.

EXAMPLE 14.   A sitting room 3 m × 4 m has a single lamp, open top, diffuser luminaire with louvred base for a tungsten lamp which gives direct illumination and is fitted 2 m above the working plane. The walls are light and the ceiling is very light. Assume a depreciation factor of 0·8. What size lamp is required for a general illuminance of 50 lux?

Area per lamp is 12 m². The room index is $\dfrac{3 \times 4}{2(3 + 4)} = 0\cdot86$.

From Table 14(c), starting at room index between 0·8 and 1·0, the fourth column of factors gives between 0·35 and 0·4, say 0·36, as the utilisation factor.

$$\text{Lumens per lamp} = \frac{50 \times 12}{0 \cdot 36 \times 0 \cdot 8} = 2083.$$

A 150 W coiled-coil lamp gives 2040 lm in Table 10 (page 130), and this is a suitable lamp to use, but having somewhat less output than necessary, it will provide slightly less light, or about 47 lux, which is near enough to requirements. This being the average value, illumination will be much higher under the luminaire and lower towards the outer parts of the room as well as 25% higher when new. For dining rooms and sitting rooms the average value of illumination should be 50–100 lux, so the above value is reasonable; but further local lighting is desirable for reading and sewing, which require 150 lux and 300 lux respectively. With larger rooms, especially long rooms, more than one luminaire is necessary for better light distribution, unless lamp-standards or wall brackets are provided, but these are often more ornamental than useful.

In drawing rooms the average value should be about 100 lux, but sufficient outlets should be provided for the use of standard or table lamps, and a piano or organ should be so illuminated that the player does not cast a shadow on the music.

In bedrooms one light should be over the dressing table, and additional outlets should be provided for bed lights and fitted basins. A separate central light for the wardrobe is often useful, especially if it contains a long mirror. Two-way switching should be employed so that the main light can be switched on at the door and off from the bedside. For the latter switch, one of the ceiling type is safer than a pendant switch.

Kitchens require 300 lux by modern standards, at the cooker, sink and work table.

EXAMPLE 15.   A kitchen, 3 metres square, is provided with two enclosed diffusing luminaires mounted on the ceiling, which is 2 m above table level. Each luminaire contains a 240 V, 60 W tungsten lamp, giving 665 lm. The ceiling and walls are of a light colour, and the maintenance factor is 0·8. Estimate the illumination provided by the original installation and recommend modern requirements.

The layout of the kitchen is given in Fig. 94. The two lights are arranged diagonally, so that the cooker, sink and fitted cabinet with its work table are illuminated. Switches are provided at each door, as shown on the plan, each controlling one lamp for convenience in switching on the light whichever entrance is used.

The room index is $\dfrac{3 \times 3}{2(3 + 3)} = 0 \cdot 75$, and the utilisation factor is

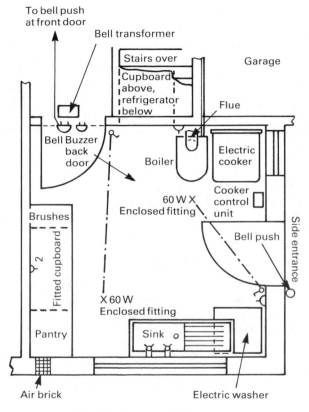

Fig. 94   Layout of kitchen

taken as 0·28, deduced from section (d) last column, Table 14, p. 141.

$$\text{Area/lamp} = \frac{3 \times 3}{2} = 4 \cdot 5 \, \text{m}^2.$$

The illuminance (lux) =

$$\frac{\text{Utilisation factor} \times \text{lumens/lamp} \times \text{Maintenance factor}}{\text{Area/lamp}} =$$

$$\frac{0 \cdot 28 \times 665 \times 0 \cdot 8}{4 \cdot 5} = 33 \, \text{lux}.$$

This figure is too low by modern standards, and at least a 200 W or 300 W tungsten lamp should be used in each luminaire, which would bring the illumination up to 110 lux or 174 lux respectively. However, few modern households maintain the high standards recommended by the C.I.B.S.E. (principally for commerce and industry), and most would probably find 150 W lamps, producing almost 80 lux, adequate, but the kitchen is most suitable for the application of fluorescent lamps, and it will be seen that, by adjusting the utilisation factor to 0·35 for one fluorescent batten luminaire in the centre of the ceiling with a bare 1200 mm, 40 W warm white lamp, 174 lux would be produced; or one 1500 mm 50 W lamp would produce 212 lux, which would be much more economical than using tungsten lamps.

Kitchens and laundry-rooms should have a light over or beside the sink, not behind the person using it. The switches should be out of reach of a person standing at the sink with wet hands, or they should be of the cord-operated ceiling type; although this is not a requirement of Regulations and the situation is not as dangerous as that which exists in a bathroom, the risk of a serious shock is nevertheless still present. Enclosed luminaires are best for damp situations, but in any case the lampholders should be properly earthed if of metal or, preferably, be of the insulated type with protective shields or skirts. Staircases and halls should have luminaires so positioned that the treads of the stairs and any changes of level can be easily seen.

The standards of lighting recommended by the C.I.B.S.E. and given here are, of course, ideal in theory, and standards are continually being raised as higher light output from lamps is obtained by research and development, but many householders cannot afford to be extravagant with increasing charges for electricity and may feel well satisfied with lower standards, especially as they may remember that standards of thirty years ago seemed quite adequate and comfortable then. This is why many homes are found with much lower levels of lighting than modern standards require. Lamp manufacturers naturally encourage higher standards and, apart from cost considerations, there is no harm in them up to daylight levels without glare. With experience of good lighting in workplaces there is the tendency to encourage the householder to use higher wattage lamps from time to time. However, we are all entitled to use our own judgement in this respect and need not worry if our home lighting is satisfactory though only half or two-thirds the recommended values since the C.I.B.S.E. state that their Code is not mandatory but based on good modern practice.

Modern decor has resulted in the liberal use of spot- and flood-lighting in the home to enhance the appearance of chosen objects or pictures, leaving other parts of the room in the shade. Sometimes a lighting track is provided along which the fittings can be moved and adjusted. The higher wattage lamps used are relatively efficient but the heat generated needs adequate ventilation for which the fittings need to be properly designed. The fashion follows shop window lighting practice and cannot be considered to be good room lighting practice because it produces excessive contrasts and makes seeing more difficult with consequent eyestrain.

It should be explained that lighting is an imprecise science because of the variety and uncertainty of reflectance values of colours and furnishings that will be provided in a room, especially in homes, so that a scientific approach in this direction is not very sound.

Emergency lighting

Although public electricity supply is extremely reliable, there are occasions when severe weather conditions put overhead lines out of action in suburban and country areas, and consumers have to resort to candles and paraffin lamps in such emergencies. There are now some good emergency electric lighting units available, which, though intended for statutory lighting requirements in certain public places, are small enough and quite suitable for a similar purpose in the house. They are simply bowl shaped or square luminaires fitted with a small battery, charger, low wattage lamps and automatic control which switches on if the mains supply fails and recharges the self-contained battery when the supply resumes. These units are described at the end of Chapter 4 (page 55).

9

Electric Space and Water Heating

Only 25% of the heat energy in a coal fire warms the room, so that only about 7 MJ (the energy in megajoules or 10^6 joules) out of 28 MJ in every kg of coal burnt are radiated out into the room. By comparison, one kilowatt-hour or one unit of electricity is a constant quantity; the quantity of heat contained in one unit is 3·6 MJ, and the efficiency of conversion to heat is 100%. Therefore all the heat given off by an electric heater is used to warm the room and a 1kW electric heater would be as good as rather more than 5 kg of coal. The installation costs of an electric heating system are usually less than those of other systems; maintenance costs are much lower, and control, cleanliness and convenience are so much better, to say nothing of the energy wastage of other fuels due to their ventilation needs, unburned fuel and limited controllability, and their contribution to environmental pollution is considerable. Thus, although electricity is the most expensive of fuels, its advantages contribute much to its popularity for heating in the home. Low rates for off-peak heating also offer savings in running costs and help to make electricity comparable with other systems of heating.

It is impossible to give comparative costs of various heating systems in the present era of inflation and lack of stability of energy costs, but the long-term outlook following the development of atomic power stations and their almost limitless low-cost fuel, while the limited supply of fossil fuels diminishes, means that electricity may well become the most attractive and economical source of domestic energy available within the present century.

Temperature control

The most important means of effecting economy with electric heating is to employ automatic temperature control. Thermostatic control is immediately responsive to temperature change and any other form of remote control; it avoids wasting electricity when the desired temperature has been reached and maintains a reasonably constant temperature, particularly where continuous or long-period heating is required. Thermostats can be provided separately as part of the wiring installation and fitted in each room, or they can be incorporated in convector-type heaters at low level to sample the cool air flowing into the heaters. In this way each heater has its own automatic controller. It should be appreciated that comfort can be obtained from a radiant heat source with an air temperature several degrees lower than would be necessary with convected heat – for example, in sunshine on a snow-capped mountain. Therefore, air thermostats controlling radiant heaters may not be so effective as when controlling convectors but will, nevertheless, serve the purpose reasonably well. Most room thermostats are either of the bimetal strip type, in which differential expansion of two dissimilar metals with rising air temperature causes contacts to open, or of the bellows type, in which the expansion of a liquid or vapour in a sensitive phial, with rising temperature, causes a bellows to operate the contacts; the phial may be remote and operate the thermostat through a length of capillary tubing. An important feature of thermostats is the temperature differential between switching on and off; this can range from ¼°C to 1°C or more, and it is obvious that the closer the differential, the less the variation of room temperature will be with the switching of the thermostat. An accelerator heater within the thermostat is often utilised to hasten the action and so reduce the differential. The switching action makes the current capacity of thermostats very limited with d.c. but ample with a.c. It is generally necessary with d.c. for the thermostat to operate a larger capacity contactor in the heating circuit, but with a.c. thermostats can handle the full circuit current up to 15 A or 20 A directly.

Thermostats should be freely exposed to air but not directly subjected to radiant heat nor should they be fitted close to the floor or in the paths of cold draughts. Warm air increases in temperature towards the ceiling, and in some rooms with convection heating and an average temperature of 16°C it has been known to vary from 10°C at floor level to 20°C at ceiling level. So, if a thermostat is set at 18°C and is placed too low, then the mean temperature of the room at the higher planes occupied by the human body will be in excess of a comfortable working temperature. A suitable position is in a fairly protected place, preferably on an outside wall, about 1·5 m above floor level (see Fig. 99, p. 159).

Internal and external thermostats can be combined with a time switch and contactor for installation control, and a number of manufacturers design and supply such control units. With thermostatic control the heat output is limited to that required by the heat losses and air temperature. Thus the electrical consumption will be the same whether the loading installed is just adequate or excessive, and a high heater loading does not involve excessive electricity consumption but can cope with wider temperature differences.

Diversity of operation of a number of thermostats reduces the maximum power demand in large installations, but this need not be taken into account for the average small house.

To calculate the heat required
The design of a heating installation with any degree of accuracy is a matter for expert and laborious calculation as it involves working out the areas of the various room surfaces through which heat is lost and multiplying each surface area by the 'U' value, (which is the amount of heat transmitted to the outside per unit area, per degree difference in temperature between inside and outside); multiplying by the temperature difference between the room temperature and the lowest outer temperature to be considered; including the air change loss (number of air changes due to ventilation per hour multiplied by the air heat factor and temperature rise); and adding together all the various heat losses to obtain the continuous heat loss per hour under the worst conditions. An allowance for abnormal exposure to weather of outer walls is also made. The 'U' values are obtained from published tables for various building constructions

Table 15 Typical heat transmission factors for some common building materials

Construction/material	Heat transmission factor Watts/m^2/°C (U value)
Solid concrete floor on ground:	* say 0.36 for full temp. difference
11″ cavity wall plastered inside:	0·96
Timber and plaster partition, or 3″ breeze partition, or 4½″ brickwork, or doors:	say 2·5
Ceiling under wood joist floor:	1·7
Windows, single glazing:	5·7

Air heat loss can be taken as 0·33W/m^3/°C/per air change/per hour.
* Higher ground temp. than for air allowed in U value.

under average normal conditions of exposure. The heat loss has then to be balanced by an equivalent heat input from the heaters.

The following example shows how the figures in Table 15 are used to calculate heat losses and the electric loading required for a typical room in a house.

EXAMPLE 16.   What size electric heater would be required for the living room of a house in a rural area which is 4 m long, 3·5 m wide and 2·5 m high, with a solid concrete floor, one long and one short external wall, 4 m² window area, a long partition to hall and stairs and a short partition to kitchen; timber/plaster ceiling. Temperature of 21°C to be maintained with 2 natural air changes per hour. Grade I central heating required, on N.H.B.C. basis.

Part	Volume or area		U value	Temp. diff.	Watts
Air	$4 \times 3·5 \times 2·5 =$	35 m³ × 2 a.c. ×	0·33	× 22 =	508
Floor	$4 \times 3·5$	14 m²	0·36	22	111
Window		4 m²	5·7	22	502
Wall	$[2·5(4 + 3·5) - 4]$	14·75 m²	0·96	22	312
Ceiling	$4 \times 3·5$	14 m²	1·7	5	119
Hall Part'n	$4 \times 2·5$	10 m²	2·5	5	125
Kit. Part'n	$3·5 \times 2·5$	8·75 m²	2·5	3	66

	1743
Add 10% for exposure to open countryside:	175
Total:	1918
or add 20% for exposure and	1743
intermittent use:	350
Total:	2093

A 2kW heater could therefore be selected, but consideration should be given to a higher loading in view of frequency of door opening into the cooler hall outside; this would make the choice of a higher loading a wise decision, and with integral switching on the heater a lower output could be obtained when necessary.

There are, however, simpler and approximate methods of ascertaining the heater loading based on average conditions and only involving a knowledge of the room dimensions. The graph given in Fig. 95 enables approximate calculations to be made for average buildings when the window glass area is not more than 6 m² per 100 m³. The height of the room being known, the watts per cubic metre

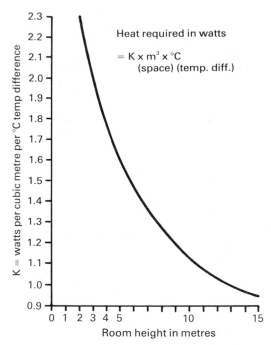

Fig. 95   Space heating graph for rooms with plastered solid brick walls (without insulation)

of space per °C temperature difference between the inside and the outside of the building will apply for any temperature differences.

Although this curve can be used to approximately assess the electrical heating load in the rooms of an average building, it is not accurate enough for designing a particular heating system for individual rooms in which, as the example shows, a ground floor living room requires more heat, with abnormal exposure and larger window area than that indicated by the curve; but other rooms in the house might require less heat.

The National House Building Council stipulates standards for heating in new houses, for which heat loss calculations must be based on an outside temperature of −1°C, and two air changes per hour also in the cases of Grades I and II.

Table 16   Temperatures of rooms in central heating of dwellings (N.H.B.C.)

Room	Minimum air temperature °C		
	Whole house heating		Background heating
Grade:	I	II	III
Living room	21	17	13
Dining room	21	17	13
Kitchen	18	13	10
Hall, landing	16	10	10
Bedrooms	16	10	10
Bathrooms	21	*10	10
W.C.	16	*10	10

Note: Grade III is intended to cover rooms if the builder provides heating for them. In all other dwellings not provided with central heating the main living room must be provided with a fixed heater having a rating equal to 42 W/m³, but minimum 2 kW.

* In Grade II these rooms may be assumed to be heated by other heated parts of the house.

Radiant and convector heaters

Space heating equipment can be divided into three main types: the radiant type, in which there is a red-hot element with a reflector, such as the popular 'electric fire'; the non-luminous radiant panel type and the convector type of heater. The true convector transfers almost all its heat to the air current passing through it, whereas the radiator transfers most of its heat directly to the objects on which its radiation is directed; this is about 70% in the case of the reflector fire heater. Types of heater that transmit heat both by convection and by radiation in varying proportions include the tubular heater, the hot water and oil-filled 'radiator', and the panel heater.

There are also combined radiators and convectors, which have the advantage of separately controlled radiant and convected heat. A visual effect of burning coal is obtained by vanes rotating over flame-coloured lamps which cause flickering reflections on the back screen as they are propelled by the rising warm air. This effect can be used without the heating elements and can also act as an indicator to show that the fire has not been switched off from the mains. This

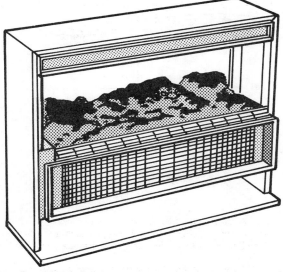

Fig. 96 Combined radiant and convector heater with coal fire effect
(*Belling & Co. Ltd.*)

type of combined heater is illustrated in Fig. 96. This model obtains
seven degrees of heat output from 0·05 kW to 2·85 kW by means of
the regulator. Radiant heaters with the coal effect but without the
convector and of similar appearance are also available.

A popular heater is the fan heater shown in Fig. 97, which
combines a heating element and fan in a very small space to blow out
a strong current of warm air for rapidly heating a room. These units
can stand on a table or on the floor or be designed for wall fixing. An
important feature of these heaters is an automatic device which
switches the heater element off if the fan stops or the airways are
obstructed when switched on because the heater cannot operate
safely without overheating if the fan is not working. Safety-guards
must be fitted to all electric fires to comply with B.S. 3456 and with
the Heating Appliances (Fireguards) Regulations 1953. Portable
heaters are usually fitted with 2 m of three-core flexible lead.

A heater designed for wall mounting is also shown in Fig. 97. Such
a fitting is suitable for the bathroom, but it must be mounted high up
on the wall out of reach of a person using the bath. The 300 W
radiant heater lamp often used in bathrooms and toilet light pendants are very limited in effectiveness because at least 500–1000

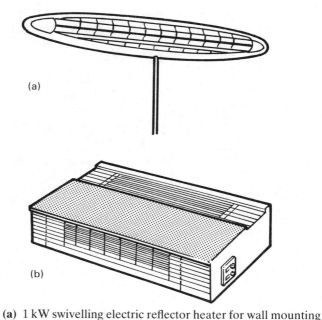

Fig. 97   **(a)** 1 kW swivelling electric reflector heater for wall mounting
(b) a 2 kW fan heater

(*Dimplex Heating Ltd.*)

watts are usually necessary for enough heat output unless one stands immediately under the lamp, the excessive brightness is unnecessary and uncomfortable, and the heat may be too great and unsafe for an ordinary flexible pendant and lampholder designed for a 60 W lamp, which usually suffers damage as a result of efforts to get the lamp out for renewal, when necessary. Properly designed overhead fittings with separate heat and light elements and switches are available but they must not be fixed within reach of a person standing in the bath and should be high enough above head level to avoid discomfort.

Fig. 98 shows a typical oil-filled panel radiator fitted with a thermostat. These heaters provide approximately half the heat output as radiation and half as convection. The heater element is immersed in an oil reservoir at the base and the oil circulates through channels in the panel. In modern buildings convector units can be concealed in walls or window recesses so long as air inlet and

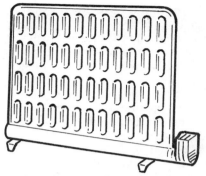

Fig. 98   1 kW oil-filled electric panel radiator and thermostat
(*Dimplex Heating Ltd.*)

outlet grilles are provided, thus presenting a smooth exterior surface. This type of heater usually employs finned tube units, the
heating element being contained in the tubes.

An electrically heated towel rail in the bathroom is a very popular
unit. This consists of oil-filled tubing with an immersion heater at
the bottom having a loading of between 70 and 250 watts depending
on the size of the towel rail. This unit should be efficiently earthed
and fixed out of reach of a person in the bath. A control switch with
pilot lamp should also be provided and fixed out of reach as well.

Location of heater

It should be understood that a convector heater will only heat the air
above its own level because the warmed air rises in the room. It must
therefore be placed as near the floor as possible, otherwise a layer of
cool air will always remain below the level of the heater. Since
radiation is always emitted in straight lines in all directions from the
element (as from the sun) and only redirected by a reflector, a
radiant heater and other reflector 'fires' can be placed at any level
desired, provided that the angle of the heater is directed towards the
part of the room where the heat is mainly required.

It is best to locate convector heaters in the coldest parts of a room,
which are generally below windows where the air, cooled by the
cold glass, starts a cold down-draught. Radiant heaters are best
placed on or against the coldest walls. Heaters that depend on a
large proportion of heat emission by convection from their surfaces

should have plenty of air space around them, especially wall-mounted panel types, which should be spaced not less than 50 mm off the wall. Panel heaters and block storage heaters which present a large surface to a cold wall behind them will lose much heat to and through the wall, therefore it is sound economy to fix a sheet of reflective and thermal insulating material to the wall behind them to minimise this heat loss.

It is most important with all types of non-luminous heaters to ensure free heat dissipation at all times, because restricted ventilation causes the heater temperature to rise until either it forces the heat to escape or a fire is started. Therefore, clothing and the like should never be placed on such heaters to air or dry. This is a most dangerous practice which has led to catastrophic fires on a number of occasions. All non-luminous heaters should have a pilot lamp or lamps, either self-contained or situated somewhere in the room (in the thermostat or control switch) to indicate when the heater is operating.

Convector heaters run at a lower temperature than the elements of an electric 'fire', consequently their elements have a longer life and require less maintenance; they also constitute less fire-risk. Convector heaters are more suitable for continuous use for space heating, as it takes time for the warm air to be circulated in contrast to the more immediate effects of radiant heating, but large air changes with ventilation could make them less economical.

Tubular heaters

This type of heater warms a room partly by radiation and partly by convection, in a similar way to the oil-filled panel heater; it consists of a hot wire element in a circular steel tube about 50 mm diameter or in an oval tube about 76 mm by 25 mm. The lengths of the tubes vary in standard sizes from 300 mm to 3·6 m and are supported by wall brackets or floor mountings. The heating element inside the tube consists of a nickel-chrome wire, which runs at black heat and is supported on mica insulators. One end of the tube is closed by a steel cap, while the other end carries a terminal assembly with two brass terminals connected to the element and a brass earthing terminal solidly connected to the tube. The standard loading is 200 watts per metre, but elements for 265 watts per metre are also made. With the lower loading the surface temperature of the tube is about 80°C, while with the higher loading the surface temperature is about 106°C. These temperatures are much too hot to touch and, unless the heaters are fixed in a safe place where they are not likely to be touched, they should be fitted with a suitable wire guard. Fig. 99 shows the arrangement of two-tier tubular heaters with connections to conduit under the floor and a thermostat in the corner of the room well away from the heaters. The heaters should be situated around the skirting under windows to minimise the effect of down-draughts and cold-air inlets. This type of heater is often used at the base of high windows and skylights for this purpose. Although the tubular heater is much used for industrial, agricultural and other applications, it has been largely superseded for house warming by the less conspicuous skirting heater and other forms of heater, because it takes up too much wall space in a room and is not very attractive in appearance.

Panel heaters and ceiling heating

There are two main types of radiant panel heating: with high-temperature panels, which work at a surface temperature of approximately 204°C; and with low-temperature panels, which operate at much lower temperatures, from about 93°C for wall panels to about 32°C for ceiling panels and 24°C for floor warming. The former type are usually fixed out of reach at high level and suspended or mounted on inclined brackets so that the radiant heat is

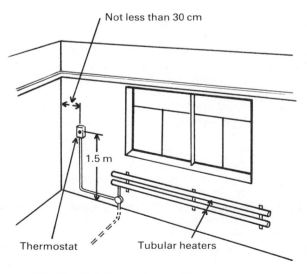

Fig. 99 Tubular heaters and thermostat

directed downwards. The latter type can be fixed against a wall or embedded in either the walls or the ceiling. Even with the lower temperatures, the backs of the panels transfer a fair amount of heat to the structure, and heat-insulating pads are desirable on outside walls behind the panels. Similarly, roofs above top-floor rooms with ceiling panels should have extra good insulation to limit the heat losses increased by the heaters themselves.

Ceiling heating is usually designed for a maximum surface temperature of about 32°C so as to avoid discomfort with too intense heat on the head. For this reason, it is unsuitable for low ceilings, and the best results are obtained with ceiling heights over 2·5 m. The limiting surface temperature necessitates low values of watts per square metre, therefore the whole ceiling area must be utilised to obtain the requisite load for the room. It is efficient, with good insulation, the heat response is fairly rapid and where there are no obstructions in the path of the radiant heat the heat distribution is excellent.

Floor heating and thermal storage

Electric floor heating became an ideal method of heating in pre-war years because of its low initial cost compared with that of other forms of central heating and its high amenity and comfort values, but it has proved too costly to use, unless a low cost per unit is available, and it cannot be regulated closely. Therefore it is only suitable for continuous heating with off-peak supplies and tariffs in which low unit rates operate during restricted hours. It is admirably suited to solid floors which become storage heaters, storing the heat input during off-peak hours and emitting heat continuously over the 24 hours.

This method of operation is quite suitable for houses generally, but a compromise scheme is frequently adopted in which the groundfloor or living rooms are provided with floor heating to provide a limited temperature rise – say, to 12°C or 15°C – as 'background' heating, keeping these rooms mildly warm continuously, and other heaters on unrestricted supply (at higher tariff rate) are used for shorter periods and in bed-rooms as required to boost to higher comfort temperatures. The additional heat required for this purpose – say, for an extra 3°C or 5°C – is very small and is well covered by a 1 kW heater in most average rooms.

The surface temperature is limited to about 24°C to avoid discomfort to the feet. The loading per square metre is therefore low, and the whole of the floor area is usually required to obtain the necessary electrical loading. The presence of furniture and floor coverings is not as a rule a very great disadvantage because, although they have an insulating effect and retard the emission of heat, the floor temperature in the covered area rises to overcome this until the required emission is produced. Floor heating is very effective and no space is taken up by heating equipment, which is a great advantage. The heating elements can be embedded in the walls in addition to or instead of the floor. Floor or wall heating for unrestricted supply requires good thermal insulation to reduce heat losses behind the elements or panels, and wood-joist floors and partition wall can be used. The heat response is fairly rapid and very good heat distribution is obtained, but heating panels embedded in walls and ceilings require special heat-resistant paints and finishes if the heated areas or lines of the elements are not to show through in time. For off-peak heating, however, the thermal storage necessarily requires a mass of heat-storage material; thus wood-joist floors cannot be used for this form of heating, which is usually only applied to solid concrete floors or to hollow tile floors with sufficient thickness (not less than 65 mm) of cement screed on top (see Fig. 100).

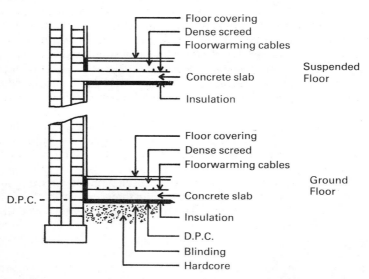

Fig. 100   Typical floor warming construction for off-peak heating installations

Thermostatic control for floor heating with unrestricted supply is, of course, essential, but with off-peak heating it is not possible to control or regulate the heat output of a heated floor artificially and natural regulation is the only means of control. In this way, the heat output depends on the difference in temperature between the floor surface and the air. Therefore, as the room temperature tends to fall with cold weather, the floor emits more heat to compensate for the heat loss, and when the temperature rises with warmer weather the heat emission from the floor is reduced accordingly. Thus the temperature conditions of the room and floor are constantly trying to balance; but the floor can never raise the temperature during the period of heat discharge in supply-restricted hours because it is constantly losing heat with falling temperature, and the temperature of the room gradually, though slightly, falls during this period. The purpose of thermostatic control in the room heating circuits is to limit the maximum room temperatures and consequently those of the floors during the off-peak period, while they are being charged with heat.

Block storage heaters

The block storage heater was a natural development from floor heating for off-peak supplies, but it has the great disadvantage of size and weight, because its heat storage blocks must have enough bulk to take in sufficient heat during the few off-peak hours to escape gradually over the whole 24 hours in order to keep the room comfortably warm. This type of heater was very popular with the advent of the simple two-rate, off-peak tariff – later known as the white meter. Economy 7 or other-named domestic off-peak tariffs offered by the Electricity Boards – in all areas; it has enabled off-peak electric heating to be installed in homes at low initial cost and with running costs low enough to be comparable with the costs of other forms of heating. It is essentially a convector heater, and in its simplest form consists of a number of blocks of cast iron or refractory material through which the heater element wires pass and to which the heat is transferred, so raising their temperature several hundred degrees. The blocks are surrounded by suitable insulating material and are enclosed in a metal case with a fusible link in circuit as a protection against overheating. In common with off-peak floor heating, this form of storage heater in its simplest form has no control over the heat output and thermostats cannot be used for this purpose. It is possible, however, to obtain this type of construction with air passages through the heater blocks and shutter regulators in the output air passage which can regulate the flow of air through the heater and the heat output to the room to a limited extent.

A modern form of this type of heater is shown in Fig. 101. It has an input control thermostat which is adjusted to suit weather conditions or outside temperature, so regulating the amount of heat stored; a room temperature sensing thermostat which adjusts the output air damper valve, so maintaining the room temperature as constant as possible; wall fixing and floor standing feet which tuck under carpeting to allow easy floor cleaning underneath. The storage core is of iron oxide/clay refractory brick with mineral insulated elements sheathed in metal (incaloy) and the insulation used is opacified silicatous aerogel and mineral wool. These materials enable the dimensions to be reduced to a minimum and the depth of space taken from the wall is only 207 mm.

The form of storage heater with the closest temperature control and most efficient performance has a fan-controlled output. In this, a maximum of insulation is used to minimise the standing fixed heat output from the heater, and a fan in the air passage forces air through the heated blocks to give a heat output that is under complete control, simply by switching the fan on or off as required. This is best accomplished automatically by a room thermostat connected in the fan circuit on unrestricted supply and, provided that the heater has enough storage capacity, the room temperature can be closely controlled or varied at will throughout the day. In

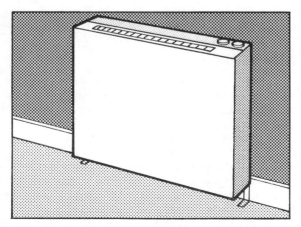

Fig. 101   Block storage heater with input and output controls
(*T.I. Creda Ltd.*)

addition to the input control on the heater the fan can have a time switch to run the fan at high speed for a short period for rapidly heating up the room when switched on in the mornings. This heater was the forerunner of the Electricaire system of central block storage heating described in the next section.

Calculating heat storage loading
The method of estimating the loading required in a storage heater is a simple calculation, since the total continuous heat output in watts for 24 hours has only to be divided by the number of off-peak hours of heat input.

Thus, assuming the maximum requirements for a particular room with a controlled output heater are 1·5 kW of heat output for 15 hours of occupation and 0·5 kW for the remaining 9 night hours each day, and the off-peak hours are 11.00 p.m. to 7.00 a.m., or 8 hours at night, then the loading required will be not less than

$$\frac{1\cdot5 \times 15 + 0\cdot5 \times 9}{8} = \frac{27}{8} = 3\cdot4 \text{ kW}.$$

The fixed (uncontrolled) standing heat output of these heaters is about 15% of the rating, so the standing heat output during the night while the output dampers or shutters are closed or the fan is switched off would cover the night requirements in this case. It will be obvious that, if the off-peak tariff includes a period of 2 or 3 hours during the day when the heaters can be boosted, the longer charging period per day will proportionally reduce the loading and size of the heaters required.

Electricaire systems

Central heating for a number of rooms on the same principle is another application of electric storage heating. These systems, known as Electricaire, employ a much larger version of the fan-controlled output heater, and the heated air is carried from a centrally placed heating unit through ducts to the various rooms in the house.

It is evident that if thermal storage systems are not well worked out and properly designed the user may find he has insufficient heat stored in the heaters for his maximum requirements and may wrongly blame the system instead of the designer – a situation not easily rectified.

For all forms of thermal storage heating with off-peak tariff, separate circuit wiring for the heaters that is under the control of a time switch is necessary, but fan circuits, if provided, must be supplied from the unrestricted supply section of the installation so that they may operate in conjunction with the thermostats during the day when the off-peak supply to the heaters is switched off. With the modern off-peak domestic tariff, however, which does not discriminate between different usages or heating equipment, such separate circuits are not essential, except that in order to get the advantage of the low off-peak rate at night the heaters must be switched off or unplugged each morning for the peak hours. However, as this procedure is inconvenient and not altogether reliable, it is better to install separate wiring and a time switch for a more satisfactory and trouble-free installation.

A system of central heating with hot water radiators can be designed or adopted for electric heating in which an electric water heater of appropriate design and capacity takes the place of the normal boiler and the water is the heat storage medium. This is called Centrelec by the Electricity Boards and it can be designed with a high degree of thermal storage so as to take advantage of off-peak electricity at low tariff rates but the water needs adequate space for storage requirements.

Solar heat and the heat pump

A look into the future reveals two developments which will no doubt benefit the householder in future years. Solar energy can be dismissed lightly because, although any energy that can be captured from the sun's rays is free, the climate in the UK is too cloudy and lacking in direct sunlight to be worth the initial cost of utilising this very attractive source of heat. An installation can be likened to a hot water radiator being exposed to the sun on the roof of a house in which the water channels are connected simply to the hot water storage tank and pipework in the house and a small pump with thermostatic control circulates the water warmed by the sun; but the heat collectors are sheets of heat absorbing material under glass with pipework attached to the underside. A very large area of these is necessary to collect enough heat for a satisfactory installation and the 2 or 3 m^2 of panels often seen on the south facing roof slope of suburban houses in this country are generally inadequate, becoming unsatisfactory and disappointing to the owners. There is no doubt, however, that in sunnier climes solar heating can produce most satisfactory results.

The heat pump, on the other hand, is an entirely different picture which does not depend on the vagaries of the weather. It is, in fact, the refrigerator principle working with a reversed purpose. Reference to Chapter 10 (p. 195) will show how a refrigerator operates – an electric motor-driven pump circulates a suitable fluid so as to extract heat from one place and dissipate it in another (inside to outside the cabinet). This is evident by feeling how hot the condenser is at the back of the refrigerator. In the heat pump a similar arrangement is designed to collect heat from the outside air or the ground at low temperatures and by compression delivering the warmed air at enhanced temperature to the inside of the house. This highly efficient piece of apparatus can produce at least three times as much energy in heat as the consumption of the drive motor. The equipment is expensive but it is not improbable that all houses will have central heating systems which incorporate heat pumps in future years.

Air conditioning

The conditioning of the air in buildings has generally been a special facility only found in the offices of large and prosperous organisations or luxury flats in city centres, but small air conditioning units are now available for use in single rooms or suites of rooms which only require connection to a local 13 A socket outlet, so that each or any particular room in a house can have air conditioning. Suitable places for the application of these units are where the outside atmosphere is unpleasant or unhealthy due to local pollution or where window opening introduces dirt, noise and security problems. In its simplest form the window fan is provided with filtration of the incoming air, and the most advanced equipment incorporates heating, cooling and filtration with temperature and humidity automatic controls. These larger equipments occupy much greater space, of course, and if standard units are not available may have to be specially designed for each installation. Air filters always have to be regularly replaced or cleaned, but apart from this there is very little maintenance required by the equipment. Economical features of this equipment can be recirculation of filtered air and arrangement of the air passages so that the incoming air absorbs heat from the outgoing air; and where cooling is required, heat pump units can extract excess heat from the room air to use for the domestic hot water supply or obtain heat from the outside air for space heating at the phenomenal high efficiency mentioned earlier in this chapter.

Where full air conditioning is provided in luxury houses containing valuable antiques and paintings, an indoor temperature of 22°C with between 40% and 60% relative humidity is generally the basis of design. Calculation of loading for air conditioning is made on a similar basis as for heating calculations but factors for heat gains from solar, lighting, power and occupant sources have to be taken into account. Consultation with air conditioning specialists is very advisable for such schemes, especially if the heat pump facilities are involved.

A system of air conditioning similar to the Electricaire system of central heating can be designed to incorporate reversed cycle operation for cooling as well as heating.

Thermal insulation in the house

The heat losses from a house depend on the temperature difference between inside and outside and the thermal conductivities of the structural materials between. As a heating system is designed to make good these losses in order to maintain the inside temperature, it is important for reasons of economy and energy saving to reduce the heat losses to a minimum by incorporating insulation materials into the structure or adding them where possible. Tests have shown that the heat losses from an average two-storey warmed dwelling are as follows:

>Roof 25%
>Walls.............................. 35%
>Windows 10%
>Floors 15%
>Ventilation or air changes 15%

Roofs and walls are therefore the areas where insulation can be most effectively applied. Roof insulation has long been the most common means of reducing losses and it is recommended that existing 25 mm thickness material be increased to 100 mm to effect a saving of 15% in the annual heating bill of an average semi-detached 3-bedroom house. It is cheap and easy to lay this between or over the joists in the roof space. Wall insulation is not so simple because so many older houses have solid brick walls and applying insulation to the inside surfaces is expensive and troublesome, but hollow or cavity walls are very easily filled from the outside with foamed or fibre insulation material pumped through small holes in the brick-work and this is well worthwhile because it can effect a saving of up to 20% of the heating bill. Such materials should be non-hygroscopic, non-combustible and without noxious fumes. Double glazing for windows is a satisfactory scheme as it reduces both window losses, noise and draughts but is fairly costly, although several do-it-yourself fitment sets are available. All methods of insulating reduce condensation by keeping internal surfaces warmer.

Electric water heaters

It is common practice to provide an electric water heater in houses with coal, gas or oil central heating systems so that hot water is available in summer time when space heating is not required and the boiler may be shut down. Consequently there are very few houses without this facility.

There are five main types of electric water heating equipment:

(a) Storage heaters of the pressure type.
(b) Storage heaters of the displacement (or non-pressure) type.
(c) Immersion heaters (including internal and external circulators).
(d) Instantaneous heaters, or geysers.
(e) Electrode boilers.

Pressure-type storage heaters are usually installed in new houses and when an existing domestic hot-water system is being converted, as they represent the best design of water heater and are the most efficient and the cheapest in the long run.

Immersion heaters are widely used to supplement existing solid-fuel or gas boiler systems for summer use and where low initial cost is of primary importance, but the efficiency of the installation may be lower.

Electric instantaneous water heaters give small quantities of hot water immediately. They have a disproportionately high electric loading and consequently involve higher wiring costs and were not originally viewed with favour by supply authorities due to their heavy intermittent loads, but with electrical development this attitude has changed. A small unit requires about 3 kW, while for a bath at least 10 kW are needed for a limited flow of water, but with no standing losses they can be economical in both water and electricity consumption.

The electrode water heater is only applicable where large quantities of hot water are required for central heating or laundry purposes, and it is not used in normal-sized houses. The principle of operation depends on the heating effect of a current passing through the water between a system of electrodes suitably arranged for the supply. This type will not be considered in any greater detail.

Pressure-type storage heater

This type of storage heater is subject to pressure due to the head of water in the cistern, which is 0·1 kg/cm² per metre head. The heater consists of an inner container of brazed copper, tinned internally where required for drinking and other domestic purposes, with an outer casing of sheet steel, the intervening space being packed with granulated cork of polyurethane foam to provide heat insulation. The smaller sizes are suitable for wall mounting, while the larger capacities are provided with feet for floor mounting. An immersion heater and a thermostat are inserted in the side of the tank at low level. In some water heaters a longer heating element is inserted almost vertically in the top of the tank and reaches to the bottom. In principle, the heated water rises to the top of the tank and increases in depth until it reaches down to the level of the thermostat, which then switches off the supply, but water below the level of the lowest part of the heater element cannot be heated. Hot water is drawn off from the top of the tank and cold water enters at the bottom. This type of water heater is intended for installations with two or more draw-off points with taps.

A modern type of automatic electric storage heater is shown in Fig. 102. This apparatus has been designed to meet the demand in flats and small kitchens where space is restricted, and the dimensions are such that the UDB-20 can be accommodated in the space under the draining-board. This location is ideal because of the short length of pipe to the sink, where the largest and most frequent demand for hot water occurs. The two sizes shown have capacities of 91 and 136 litres respectively and two separate immersion apparatus plates are provided. The one near the top consists of one 1000 W element controlled by its own thermostat, while the bottom plate has one element of 2000 W controlled by a second thermostat. The top element will provide a constant supply of 32 litres of hot water sufficient for washing-up and cleaning purposes, while the bottom element can be switched on when baths are required. All the 1000 W elements are interchangeable and withdrawable when the heater is full; the internal wiring is complete, so that the electrical fitting is confined to the necessary connections to a double-pole switch and the provision of a good earth connection.

The arrangement of this water heater for an independent service is shown in Fig. 103 (page 172), in which the heater is directly fed from the house cold water storage tank. This system works at a high all-round efficiency, and it is estimated that, with electricity at 6p per unit, ample hot water can be provided for a family of five for less than £4.00 per week. With the Economy 7 tariff and adequate

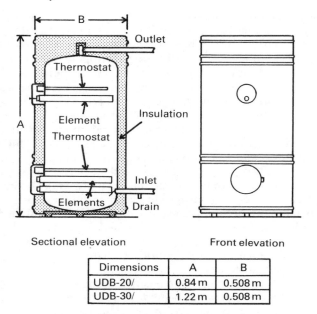

Sectional elevation   Front elevation

Dimensions	A	B
UDB-20/	0.84 m	0.508 m
UDB-30/	1.22 m	0.508 m

Fig. 102   Pressure-type electric storage heater
(*Heatrae-Sadia Heating*)

storage tank capacity for night heating, even greater economy can
be achieved. The usual arrangement with this tariff is for the bottom
heater to heat the water at night and the top heater to be available at
any time for heating 30–40 litres of water quickly if the contents of
the tank have been drawn off and no hot water remains. This is
effected by a special control unit with clock dial and push-button
timed to operate for a period up to one hour before switching off
automatically and is fitted near the tank. Similar water heaters,
generally with a single heating element, are commonly installed in
airing cupboards on upper floors where the heat loss is useful. Other
types of water heater are available with self-contained balltanks for
direct connection to the main water supply; but it should be
remembered that as the flow of water from a tap depends on the
height of the balltank above the tap (head of water) the position of
this type of heater must be as high as possible in the house or flat.
The economical importance of thermal insulation or lagging on hot
water pipework cannot be over-emphasised as this is so often

neglected and consequently results in appreciable excessive heat losses.

For normal domestic demands in a medium-sized house, a water heater smaller than 100 litres capacity should not be used unless the requirements are known to be very moderate, as barely two small baths in succession can be provided with this capacity. The N.H.B.C. stipulate 115 litres at a temperature of 60°C with full recovery in two hours as the minimum hot water storage capacity for an average family house. This would require a loading of 7·6 kW with an unrestricted supply. Table 17 on page 175 gives some idea of the average family consumption of hot water, from which it is possible to work out the storage capacity required, provided that a temperature recovery rate of 2 minutes per 1 kW heater element per litre of water used is taken into account between draw-off times.

When hot water storage is required at off-peak tariffs, it is important to calculate the amount of hot water required per day, since it is not possible to heat up more water in restricted hours when all the hot water has been used unless means are provided to switch over to the daytime supply at the higher peak-hour tariff rate as an emergency measure or a separate unit is installed for this purpose. A minimum capacity of about 200 litres is necessary for off-peak storage in the average small household.

Open-outlet types

Self-contained water heaters from 6 to 68 litres capacity to supply one point are operated on the displacement or non-pressure principle and *must be provided with an open outlet* so that pressure cannot build up within the container, which would otherwise explode with the resulting water pressure for which it is not designed. Therefore, a tap must not be fitted to the outlet. When the inlet control tap is turned on, the entering cold water displaces an equivalent volume of hot water through the open outlet spout into the washbasin or sink. Thermostatic control is provided, and the electrical and water connections are quite simple; for the 6, 10 and 13·8 litre sizes direct connection may be made to the water main; these are the usual domestic sizes. The electrical loading is generally 1 kW but where larger than normal quantities of hot water are required 3 kW elements are available which provide up to 45 litres of hot water per hour. Smaller units of this type for wall fixing are available for bringing water to the boil instead of using a portable kettle; this is very helpful for disabled people. This type of water heater is shown in Fig. 104 (page 173).

The small instantaneous water heater for washbasins is common

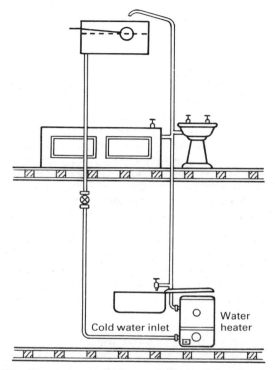

Fig. 103   Pressure water heater and pipework for three outlets

for hand-washing in lavatories; it is usually fitted with a spray nozzle on the swivel outlet, and with a 3 kW loading it delivers about 1·4 litres per minute at 38°C. It is an open-outlet or non-pressure type of water heater and may be connected to the water mains or to a cistern supply. A 7 kW shower unit (instantaneous type) is also shown in Fig. 104.

It is essential, before installing any type of water heater, to see that it is suitable for the hardness and acidity of the local water, and that the water supply authority's regulations are not infringed.

Immersion heaters

There are several different arrangements of heating elements that can be fitted to existing hot-water tanks. They usually consist of a mineral-insulated, copper-sheathed, tubular element and an

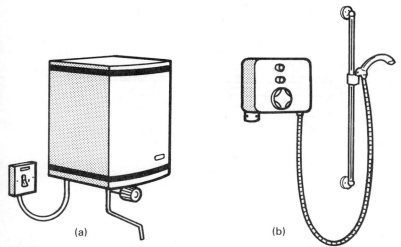

Fig. 104   (a) 7 litre, 1 kW open outlet water heater
            (b) 7 kW bathroom shower unit

(Heatrae-Sadia Heating Ltd.)

adjustable or fixed temperature thermostat in a separate tube fitted into a terminal head, with cover and screw boss, ready to insert into a suitable flange mounting, which must be provided at low level in the tank. Such a unit is shown in Fig. 105(a). This is a non-withdrawable type, because it cannot be taken out unless the tank is empty. Withdrawable immersion heater elements and thermostats are contained in hollow tubes fitted to the head which enable them to be withdrawn with the tank full of water.

Immersion heater elements suffer from furring, just as a kettle does, and they are also corroded by hard water, so the element sheaths are made of metals more immune than copper to these effects where the water is cuprosolvent. Plating, monel metal and stainless steel have been used, but the latest metal to be adopted is titanium which has proved most satisfactory, but, in common with other special metals, it is expensive, but it has been found that these metals have about double the working life of copper sheaths in some corrosive waters. The method of dealing with hard water is to set thermostats at a lower temperature than with soft waters to reduce the furring or in extreme cases to provide an automatic softener at the intake water main position of the house.

Fig. 105(b) shows a dual immersion heater which is fixed verti-

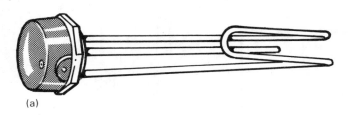

(a)

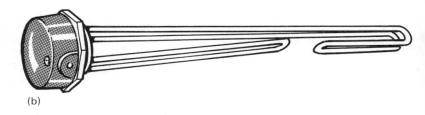

(b)

Fig. 105   **(a)** 3 kW immersion heater and thermostat for horizontal fixing
         (b) 2 kW/3 kW dual immersion heater for vertical fixing in cylinder.

(I.M.I. Santon Ltd.)

cally or almost so in the top of the tank or cylinder; the short 2 kW element heats a small quantity of water at the top of the cylinder for sink use, and the long 3 kW element heats the total capacity of water from the bottom for baths and either part can be switched on as required. This is an economic arrangement for unrestricted electricity supply. Where off-peak electricity at night rates is used, the capacity of the storage cylinder needs to be sufficient for the whole day's supply (which can be calculated from Table 17 (page 175)) and will probably be at least twice the capacity which would otherwise be required without night storage.

Table 17 Hot water used in an average household

Use	Litres required	Required temperature	Litres required	
		°C	60°C	71°C
Hot bath	114	44	74	61
Hand wash	4·5	44	3·1	2·4
Dishwashing	4·5/meal	60	4·5	3·7
House cleaning	9/day	60	9·9	7·5
Laundry	45/week	60	45·5	37

Note: The above figures assume a yearly average cold water temperature of 12°C.

Conversion of existing hot-water systems

The success of a conversion depends on the condition and layout of the existing pipework, because badly designed hot-water supply pipework can cause expensive waste of heat, with unwanted water circulating currents in the pipes, which cannot be afforded with electric immersion heaters. Similarly, when hot-water radiators for space heating are connected to solid fuel boiler systems, it is both inefficient and uneconomical to allow the electrically heated water to circulate in them. Expert advice should be obtained before deciding on such an installation.

Settings for thermostats

Although less water can be stored at higher than at lower temperatures for mixing with cold water to obtain the same quantity of water at the required usable temperature, it is not advisable to set thermostats too high, since scale deposit or furring of the immersion element becomes serious at high temperatures. In addition, the standing heat loss from the water cylinder may become excessive.

With soft water, and up to about 12° hardness, the thermostat may be set to interrupt the supply at a water temperature of 80°C;

Table 18 Temperature equivalents (Fahrenheit and Centigrade)

	°F	°C
Boiling water temperature	212	100
Scalding temperature	158	70
Washing-up dishes, temperature	140	60
Bath, average temperature	104	40

with medium hardness, at 70°C; and with hardness over 20° the thermostat should not be set higher than 60°C.

Water-heating data

With water heating we are able to calculate the power required with a greater degree of accuracy than with space heating. We know that one unit of electricity (1 kWh) = 3·6 MJ and that 1 litre of water (1 kg) requires 4180 J to raise the temperature 1°C. Therefore, the consumption of electricity will be:

$$\text{Number of units (kWh)} = \frac{\text{Litres} \times \text{Temperature rise °C}}{860 \times \text{Efficiency}}.$$

Allowing for losses, the efficiency varies from about 92% with good lagging to less than 80% for unlagged tanks. From the above equation the size of heater required to heat a given quantity of water in a given time may be ascertained. Assuming, for easy calculation, an efficiency of 87%, we obtain 750 in the denominator; then

$$\text{Size of heater in kW} = \frac{\text{Litres} \times \text{Temp. rise °C}}{\text{Time in hours} \times 750}$$

$$\text{and litres} = \frac{\text{kW} \times \text{Time in hours} \times 750}{\text{Temp. rise °C}}.$$

A room temperature of 60°F is approximately 15°C.

Cold-water inlet temperatures may vary from 5° to 21°C, depending on the locality and the time of year, but 5°C is used for water-heating calculations in mild winter climates.

Approximate capacity of tanks

Cylindrical tank = (Diameter)^2 × Height × 785·4 litres. Rectangular tank = Length × Breadth × Height × 1000 litres. All the above tank dimensions are in metres.

As a rough rule, 1 unit of electricity will boil about 7 litres of water starting from cold in a kettle.

EXAMPLE 17.   A cylindrical tank is 300 mm diameter and 650 mm high. How many litres will it hold, and what size of immersion heater should be fitted if all the water is to be heated from 5°C to 70°C in 2 hours?

$$\text{Cylindrical tank} = 0·3^2 \times 0·65 \times 785·4 = 46 \text{ litres}$$

$$\text{Size of heater} = \frac{46 \times (70 - 5)}{2 \times 750} = 2 \text{ kW}.$$

10

Domestic Appliances and the Caravan

Domestic electrical appliances play an increasingly important part in the house in the form of labour-saving devices from food preparation to 'do-it-yourself' work and Fig. 106 gives some indication of the wide variety of domestic electrical equipment to be found in the modern home. The important things to look for in domestic electrical equipment are safety and reliability. Modern plastics enable strong insulating enclosures to be made for some appliances but others need stronger construction. Where exposed metalwork is unavoidable in an appliance it must be effectively earthed for safety.

Small motorised appliances

Many of the small motorised appliances used in the home are well within 720 watts loading and should therefore be protected by 3 A fuses in their associated 13 A plugs. They are generally of all-insulated or double-insulated construction (i.e. the enclosures are of insulating material as well as the covering of live wires and other conducting parts inside). They are extremely safe to use because there is no exposed metalwork that can become 'live' in the event of an electrical fault. They do not therefore need to be earthed and their flexible leads do not incorporate an earthing conductor; the earthing terminal in a 13 A plug is left unused. Plastic materials used for such appliances are usually thermoplastics which soften and lose strength with extreme heat – this material cannot therefore be used for irons, toasters or any other heat-producing equipment. Such appliances must be constructed of metal which makes earthing essential.

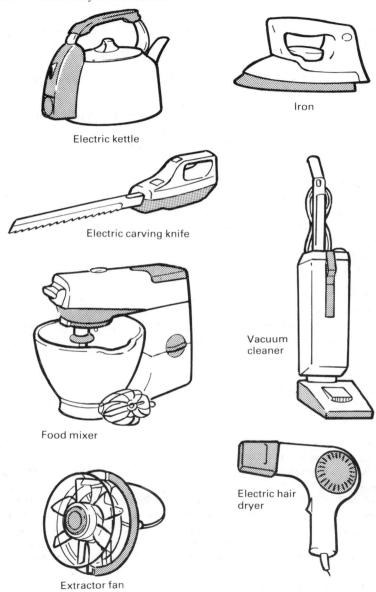

Electric kettle

Iron

Electric carving knife

Food mixer

Vacuum cleaner

Extractor fan

Electric hair dryer

Fig. 106 A wide variety of domestic electrical appliances is now available, some of which are shown above

Food processor
The food processor or food preparation appliance is perhaps the most popular unit to be found in the kitchen; it has a central powered head with speed regulator and a number of separate attachments for mixing, mincing, grinding, shredding, blending etc., even potato peeling. The loading of this unit is about 300 W. There are also separate units available for each or several of these functions with loadings up to about 150 W and about 300 W for coffee grinders and can openers. The knife sharpener is a particularly useful item which incorporates a very small emery wheel and has a loading of only 12 W.

Waste disposal unit
This is a powered crusher of waste foodstuffs, etc., which is fitted in the sink drainpipe and can crush 50 kg of waste matter to pass through into the drain with a consumption of only one unit of electricity. But the user must be careful not to drop in any metal objects such as spoons, as they will be damaged or may damage the unit itself if they enter it from the sink.

Sewing machines
The electric sewing machine has a long history of usefulness in the home. The modern machine is a complex mechanism and deserves care and attention in its use and maintenance. If there are exposed metal parts it should be properly earthed. Conversion of an old sewing machine to electric power is easily made by attaching a clip-on motor drive with rubber belt, foot control pedal and clip-on lamp to illuminate the stitching action, but it is important to ensure that the flexible leads are not entangled and damaged. The sewing machine has a loading of about 100 W or less.

Other small appliances
With the introduction of the electrically powered toothbrush and carving knife it was to be wondered how much further electrification in the home could go. Many of these small appliances are designed to operate with connections to the mains supply with almost negligible electricity consumption, and some are provided with an alternative supply in the form of a self-contained battery of the dry or rechargeable type. The latest dry batteries are alkaline cells which have a much longer life and better output than the leclanché cells of earlier batteries and are more suitable for very small appliances. The power from a dry battery is extremely limited, however, and appliances with accumulator batteries need frequent recharging from the mains.

Garden appliances

The lawnmower and the hedge trimmer are cases in point because battery driven machines are attractive in order to avoid the use of very long flexible leads in large gardens, but one must be wary of buying the larger of such appliances and expecting more power than is available from batteries. However, it is much better economically in the long run to use mains operated equipment. The best way to use a mains operated lawnmower is to work backwards and forwards starting near the house and working towards the farthest end, turning the mower at the end of each run so that the flex is left in snake fashion on the ground to avoid being coiled at any point. (The same principle applies to the use of a vacuum cleaner in the house.) Winding a long flex in figure-eight fashion avoids coiling and kinking and therefore lengthens the life of the flexible lead. An extension reel is most useful where an appliance has to be used at both short and long distances from a socket outlet. If outside the house, the socket outlet should be of a watertight type and controlled by an RCCB. (See also the requirements for such a point in Chapter 7 (page 113).

Electric blankets

The electric blanket consists of a spiral insulated element wound on a tough flexible core distributed over the blanket area and sandwiched between the blanket material. A switch with a neon lamp indicator is usually fitted in the flexible lead, and some types have thermostat contacts in the blanket itself to prevent overheating. The loadings are usually about 50 W for single bed size and 100 W for double bed size. Electric blankets should be inspected and tested, and any necessary repairs carried out, before use each winter period to ensure that they are in good condition. The dangers lie in unintentionally leaving them switched on during the day and in piling clothing on top of the bed while they are switched on, which can result in overheating and fires. There are two types available: the underblanket and the overblanket. The electric blanket should never be used as an underblanket unless it is suitably controlled by a thermostat or time switch; as an overblanket, it is less liable to become overheated. But each is made for a specific purpose with appropriate control unit attached. Underblankets should always be switched off before getting into bed, and when stored both types should never be folded in such a way as to cause them to be creased. In use they must be kept dry.

Electric cookers

This section will be confined to domestic types of cookers, and the electrical and technical aspects of them. The uniformity of the amount of heat provided from one unit (kWh) and the consistent results that can be obtained are the greatest advantages of the electric cooker, besides its general cleanliness and the absence of fumes.

The desired oven temperature is obtained by setting the control dial at the requisite temperature, which is then maintained within close limits, as shown in Fig. 107.

Simple clockwork timers or electric timers are a common feature; they simply cause a buzzer to sound when the period for which they have been set has expired. Automatic oven timers are designed to control the oven switching for predetermined periods at some time ahead, so that the cook can set the timer, go out for the day and return to a cooked meal. These timers are an example of the ordinary time switch, with adjustable cams on discs, driven by an electric synchronous clock, which close and open switch contacts for the oven, the temperature having been pre-set and controlled by the oven thermostat. Temperature regulators, of the bimetallic thermal intermittent switching type heated by the element current, are used to control hotplates.

'Plug-in' elements are provided in the majority of cookers and the spaces between the oven walls are lined with heat-insulating material. Such additional features as foot pedals for door opening, high-level grills and rotisseries, indicator lights, heated plate drawers or cupboards, oven fan, reversible oven doors, height adjustment, castors or rollers for easy removal and an interior glass

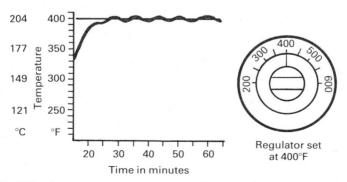

Fig. 107   Oven temperature control by thermostat-type regulator

door to view the contents of the oven without the risk of cooling draughts are provided on some models; a common feature now is the self-cleaning oven, in which extra heat over a period removes the scale resulting from cooking processes. It is also possible to obtain cookers with the oven at different levels or even as a separate unit, so that a group of hotplates, the oven and the control panel can

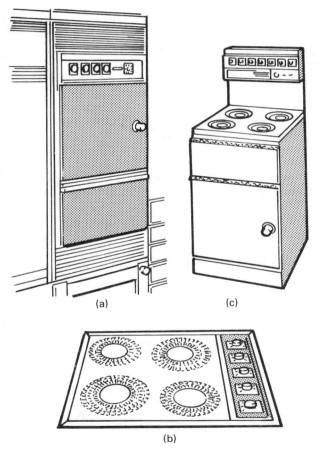

(a)

(c)

(b)

Fig. 108 (a) Flush mounting ovens and controls in kitchen cabinet
(b) Ceramic hob with control switches for mounting into a kitchen work-top
(c) Typical family-size free-standing cooker

be fitted separately or into a continuous work-top or other kitchen cabinet. Fig. 108(c) shows a typical free-standing family cooker, and (a) and (b) built-in versions for specially designed kitchens.

In considering circuit loadings, the maximum load that is likely to be on at one time can be taken to vary from two-thirds for an 8 kW cooker to one-half for a 15 kW cooker approximately. Trends in recent years have been towards higher loadings, so that, whereas in the past a 7 kW family cooker with a 30 ampère supply circuit was considered normal to provide for in a two- or three-bedroom house or flat, it is now necessary to allow for cookers with a total loading up to about 15 kW, or more. Although cookers vary so much in loading, the cooking requirements of an average family do not vary to the same extent. Even if a cooker that is much too large for normal requirements is installed, the maximum load taken is not likely to exceed that taken by a smaller cooker by very much. However, regulation requirements allow for diversity in the use of the various parts of a cooker such that a 30 ampère circuit may still be installed for any cooker with a loading up to 14 kW. Every cooker must be controlled by a separate switch placed within 2 m of the appliance but two stationary cookers within one room may be controlled by a single isolating switch provided they are both within 2 m of the switch.

There are two types of hob and hot plate: the solid plate type, which presents a smooth unbroken ceramic surface over the whole hob area, which facilitates cleaning, and from which the heat is transferred from elements below by conduction to the utensil on top: and the exposed radiant type, with a reflector plate or drip tray on the underside, in which the heat transfer is by radiation and conduction where the element is in contact with the utensil. With the former type, utensils having flat, machined bottoms that cannot buckle obtain maximum heat transmission, while with the latter type perfect contact is not essential. Grilling is done with similar elements above the grill-pan by downward radiation.

The radiant type element consists of a nickel-chrome heating spiral enclosed in a steel tube and insulated therefrom by magnesium oxide. This type of element has rapid heating response, is efficient and has loadings to bring a pint of water or milk to the boil in about 4 minutes. The loading of an 180 mm boiling ring is about 2000 W, while the 150 mm size takes about 1650 W. The grilling element generally has a loading from 1000 W to 2750 W, depending on the size. Boiling plates and grill elements are separately provided with energy regulator controls, by means of which the heat from the elements can be varied over a wide range from 'simmer' to full heat.

This form of regulator was mentioned earlier in this section. A number of variations of the so-called 'breakfast cooker' are made, from the single or twin hotplate unit to combinations, with a grill and oven below, called a boiler-griller. Fig. 109 shows a typical small 'breakfast cooker' on a floor stand which can be plugged into a 13 A or 15 A socket outlet. The average consumption of electric cookers has been found to be about 1 unit per person per day for all meals.

Microwave oven
To those who are attracted by high speed cooking regardless of first cost, the microwave cooker will have considerable appeal, and

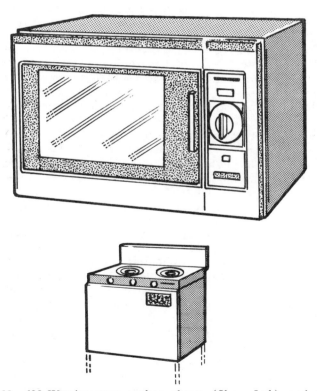

Fig. 109   600 W microwave cooker, above, (*Sharp Ltd.*), and a small 'breakfast' cooker on floor stand, with two radiant hotplates and oven/grill below (*Belling Ltd.*)

although it also has a high cooking efficiency, using about one unit an hour but cooking in minutes rather than hours, it is questionable whether the current high price justifies the low operating costs, with an initial outlay of almost £300, on grounds of economy. Cooking heat in the foodstuff to be cooked is produced by high frequency electromagnetic radiation directed from a magnetron tube over the oven which induces molecular movement in the moisture content of the food; this converts to heat within the foodstuff, without heating the container or the cooker itself, to a depth of over 60 mm, so a turntable is provided inside the oven which can be rotated to ensure even heating throughout. The turntable can be automatically power driven.

The size generally available has a capacity of about 15 litres (about a third to half a cubic foot) of usable oven space and the loadings are about 600 to 1000 W using up to about one unit an hour, so it can be plugged into a 13 A socket outlet. A larger unit of over 2 kW loading is also available. A timer, selector switch and an indicator light are fitted in the control panel at the side of the oven. The oven can be used for rapid defrosting of food taken directly from a freezer. There are certain cooking limitations; shortcrust pastry will not crisp and tougher cuts of meat require longer periods of cooking to become tender; and metal objects must not be placed in the oven, although some plastic containers are not affected by the radiation. The cooker shown in Fig. 109 is comparatively light and portable, and safety is preserved by its automatically switching off whenever the oven door is opened. The radiation frequency is about 2500 MHz.

Connection of cookers

This should be carried out by a competent installation contractor or by the supply authority's employees. The cooker circuit must be terminated in a control unit with a double-pole main switch, so that the cooker can be completely disconnected from the supply. A three-pin socket may be included in the cooker control unit for an electric kettle. A typical cooker control unit, with neon pilot lamps to indicate when a switch is on, is shown in Fig. 110. In places or installations where the supply capacity or maximum load must be restricted, overloading is prevented by having a control switch of the changeover type, so that either the cooker or the washing machine (alternatively, the immersion heater) may be in circuit but not both together, but this type of control unit is seldom necessary in modern installations. The auxiliary socket outlet may be 13 A or 15 A, but in the latter case a fuse unit for the socket is incorporated in

Fig. 110   Cooker control unit with pilot lamps and 45 A cooker switch, surface insulated type

(*M.K. Electric Ltd.*)

the control unit. Cooker control units should comply with B.S. 4177 and are available in surface or flush, all-insulated or metal-case construction. The cooker is connected to a separate circuit in the house installation by a flexible multicore cable or single-core cables in a flexible tube from the control unit. In flush wiring a separate connector box in the wall below the control unit may be used. This is shown in Fig. 111. It is essential that the cooker and its control switch are properly earthed.

Fig. 111   Flush connector unit (box not shown) for lead to cooker

(*M.K. Electric Ltd.*)

Electric kettles

An electric kettle is a useful adjunct to an electric cooker, as the use of a hotplate for boiling water is less efficient and takes up cooking space. Electric kettles have an internal immersion element that gives out all its heat to the water; this is the highly efficient feature of electric water heating. The immersion element consists of resistance wire coiled in mineral insulation within a metal tube and terminating in the connector assembly, so shaped that it can be passed into the kettle through the lid opening and the connector can be passed out again through a hole in the back, where a watertight joint is made to the kettle with rubber and fibre joint rings and a screwed outer connector casing. A safety device is usually incorporated to ensure interruption of the supply if the kettle boils dry. This may be in the form of an ejector rod which pushes the removable part of the connector off the kettle, when a fusible joint attached to the kettle element releases a spring, or it may be a thermal cut-out which breaks a contact in the heater circuit, usually by means of a bimetal strip, if overheated. Kettles should always be fitted with three-core flexible cords and three-pin plugs so that they are effectively earthed. The fixed kettle connector must always be in the form of shrouded pins and the loose flexible cord connector should be in the form of sockets, otherwise exposed live pins on this end of the flexible cord would be very dangerous. Some kettles have a thermostatic device that automatically switches the element off when boiling point is reached. Electric kettles have loadings up to 5 kW in both 1·7 and 2·3 litre capacities and, of course, the higher the loading the sooner the kettle will boil; a 1·7 litre 2 kW kettle will boil in about five minutes and all kettles will boil about 6 litres for one unit of electricity.

Electric irons, toasters and other cooking appliances

A well-balanced iron of the correct shape and weight, with a handle shaped to minimise fatigue, should be used. The most common source of trouble is with the flexible cord, as this suffers from flexing more than any other appliance lead and is liable to be twisted and to become worn with use; therefore, the flexible cord should be examined frequently and replaced when necessary. A good-quality, three-core, unkinkable flexible cord is best. In use the iron must not be left flat on the table, or it will burn right through the wood, and some fires have been caused by this happening. Ironing boards should have a raised metal grid over an asbestos pad to rest the iron on and avoid this danger. Most irons have thermostatic control with a regulator to adjust the temperature to suit various fabrics being ironed and a pilot light to indicate when the required temperature has been reached. An additional facility to save damping certain fabrics before ironing is provided by the steam iron. Steam irons are made so that a small quantity of water is put in an opening with a spring stopper on top of the iron; when hot, steam issues through grooves in the face of the iron to dampen the fabric being ironed.

The loading of electric irons varies between about 500 W and 750 W but the thermostat reduces the average loading to a much lower figure; one unit of electricity will usually provide for over two hours use of the iron.

The electric toaster is a well established appliance in the kitchen and is more economical than using the grill of the cooker for making toast; it is normally provided with a thermostat release and regulator which allows a spring to push the toast out of the toaster when the desired amount of browning has been reached, and at the same time the elements are switched off automatically. Loadings are between 500 W and 1000 W and since about seventy pieces of toast can be done for one unit of electricity this must be the most economical method of making toast.

It is most important for irons and toasters to be properly earthed for safety as previously mentioned.

The electric warming tray for food containers or plates has long been available. The modern development of this unit is the wheeled, heated cabinet and a smaller version which stands on the sideboard. Several food dishes and plates or servings ready on plates can be kept warm in them for a considerable period and this can save much time and trouble at meal times. Loadings are about 500 W to 1000 W according to the size of the unit.

A very economical appliance is the slow cooking pot which consists of an earthenware casserole of about 3 litre capacity with an inbuilt or wrap-around heating element of less than 200 W loading with control adjustment to regulate the heat to lower values. It cooks the contents over a long period of hours for a unit or two while the owner is out during the day, or it could even operate at night at the cheaper rate for electricity.

Several other individual cooking appliances are now available such as the deep fat fryer, frypan, infra-red grill, egg boiler and the rotisserie and grill unit, which are alternatives to the various parts of the normal cooker designed to carry out their purposes in a more efficient manner and more economically than the cooker itself with separate utensils. The disadvantages of these are the high cost and extra space required to accommodate them in the kitchen, but they may be useful to those without a normal cooker.

Where ventilation in the kitchen cannot get rid of cooking smells a good investment might be in a cooker hood which filters the air rising from the cooker; it has a loading of 100 W and, of course, the filter needs periodic cleaning. An extractor fan fitted in a window pane has negligible consumption and may be equally effective. This should be controlled by a ceiling switch nearby.

The teamaker works like an electric kettle but is of smaller capacity and loading; the steam may activate a mechanism to immerse the tea in the boiling water automatically. Various designs of coffee-making appliances are also available arranged to percolate at a controlled temperature and then keep the coffee container on a hotplate for immediate use.

Vacuum cleaners

The working parts of a vacuum cleaner consist of a high-speed motor with a shaft extension at one end, to which is attached a fan. The motor is of the universal type described in chapter 3 and may run as high as 10,000 revolutions per minute. In the modern vacuum cleaner the dust bag is inside a canister, which allows the suction of the fan to draw the dust into the bag, the air being discharged over or through the motor or used at that end for 'blowing' instead of 'suction'. It is common to have also a rotating brush driven by a rubber belt from the motor shaft. The most frequent troubles are blockage of the vacuum hose and damage or kinking of the trailing flexible lead, which should be frequently inspected for abrasion or damage to the insulation. The motor brushes require periodical renewal; and, if the motor will not start, a test should be made to see if there is a break in the series circuit. This might be due to a broken wire or to bad contact at the switch, brushes or terminals.

There are two main types of vacuum cleaner. One is similar in shape to the manual carpet sweeper, with the dust bag and container forming part of the handle and with a power driven brush in the wheeled base; the other is a cylinder shape with wheels or skids to slide over the floor and a length of extension hose with a variety of cleaning tools. However, it does not have a powered brush.

As the first type is also usually provided with a hose attachment and tools it is the most useful type to have, particularly where there are large areas of carpet to be cleaned. Three or four hours use will consume about one unit of electricity.

Some cleaners have an internal drum for automatically rewinding the flexible lead but the upright type usually has a pair of hooks for storing the flex. The flexible leads are necessarily very long and require much care in handling and it is wise to wind them in figure 8 fashion on the cleaner or hooks provided rather than in a straight circular coil as emphasised earlier in this chapter.

Laundering appliances

The modern washing machine is one of the most advanced pieces of domestic equipment and it is certainly the most labour saving of appliances. The washing load is put into a revolving stainless steel drum which is accessible through a front or top hinged lid and turns the washing over slowly or spin-dries at high speed; a pump fills or empties the machine with water, charged with detergent, at the correct temperature and controlled by a heater and thermostat. A large range of washing processes is automatically programmed by an electronic controller to suit various fabrics, from start to finish.

Electric washing machines are constructed of welded sheet steel with a stove-enamel finish. In general, the simpler types usually have a rotary agitator or gyrator in a tank, a pump and hose connections – one to the hot water tap and the other to waste at the sink. A heater for the water can be provided as an alternative to a hot-water supply from the tap. The pump empties the tank through a hose in a few minutes. Washers such as that shown in Fig. 112 (page 191) are made to take up to 4 kg of clothes and larger machines can also be obtained. The motors are about 350 W (main) and 110 W (pump) single phase induction type with a heater of about 2·5 kW. The maximum overall load is about 3 kW. By means of electronic, integrated circuit, computer-type controls, sixteen different wash programmes for eight different kinds of fabric can be processed automatically and left to complete the washing without attention. Some machines can operate up to twenty programmes. Automatic safety devices are also usually incorporated for protection against mal-operation. To obtain trouble-free service over many years and to maintain the washing machine in good condition it should be regularly inspected by an authorised dealer at least once a year, besides carrying out the simple maintenance prescribed on the instruction sheet.

Modern automatic washing machines spin-dry at high speed as the last process in each programme and usually have permanent fixed plumbing connections for water supply and draining but can be provided with hoses for tap connection and drainage to sink if required. The automatic washing machine can be expected to carry

Fig. 112 Modern washing machine with automatic control of sixteen wash
programmes

(*Hoover Ltd.*)

out the weekly family wash for four people with an electricity
consumption of about 9 units.

Spin driers
Spin driers have internal perforated containers which run at a speed
of about 1000 rev/min and throw the water out of the wet contents

by centrifugal force in a few minutes, leaving the fabrics almost dry enough for ironing. Capacities up to 2·7 kg of clothes are available. Tumble driers are also made, in which the clothing is tumbled over at much lower speeds with heated air flowing through the fabrics, and this gets clothing much drier than the spin drier – ready to iron or wear. They usually have a glass front-opening door for inspection. Automatic control may be provided to regulate the period of drying, and all models have switch-interlocked front or top lids to stop the machine if the lid is opened during operation. A spin drier consumes about a quarter unit per hour and a tumble drier about two units per hour of use but, of course, they are used for much shorter periods on each occasion.

Dishwasher
This piece of equipment can take up almost as much space as a washing machine, but smaller versions are available which can stand on a draining board by the sink. They can be attached to hot and cold taps by rubber hoses or permanently plumbed to the water pipework in the same way as washing machines are connected. The crockery is placed in a sectional wire cage and is cleaned by hot or cold water being sprayed on to it at high pressure. The electrical technical aspects are similar to those of the washing machine and the electricity consumption averages about one unit to clean dinner dishes for four people. This, again, is a machine that requires proper earthing, especially if the water connections are rubber hoses and the unit is not solidly plumbed into the water pipework.

Rotary ironer
The most popular type of rotary ironer is provided with a 500 mm or 650 mm long padded roller, rotated by a small motor, and an electrically heated pressure shoe of stainless steel. The smaller size is more in demand as it is much lighter for handling and can stand on a bench or table. The ironing shoe has a loading of 1000 to 1500 W according to size, usually with an energy regulator to control the iron temperature similar to those used on cookers.

The domestic refrigerator

A refrigerator must have efficient thermal insulation; the door must be tight fitting so that there is no air leakage; and the internal air circulation must allow the air to flow easily over the cooling surfaces. Easy cleaning is essential, and the parts containing the refrigerant must be of robust construction so that there is no

Fig. 113 Typical modern compressor refrigerator with separate freezer
compartment at top (4 star)

(Electrolux Ltd.)

possibility of the escape of the liquid or gaseous refrigerant. Refrigerators are graded to suit the length of time that frozen food can
be stored in them, i.e. one star for one week, two stars for up to one
month and three stars for up to three months. Defrosting of the ice
around the freezing compartment should be done weekly if not
done automatically, as is the case with the latest models available.
Average consumption is about one unit a day.

The electrical refrigerator can be operated by either the mechanical compression system, or the non-mechanical absorption
system. In the former a motor-driven compressor is used, while in
the latter a heater is used to circulate the refrigerant round the
system.

Absorption refrigerators are made for operation by electricity, gas or oil. The inherent efficiency of the absorption system is much lower than that of the compression system; but they are practically noiseless, whereas the motor and compressor are not. Modern domestic refrigerators of the compressor type are so quiet, however, that this is not usually a problem and the majority sold are of this kind.

Principle of operation

When a liquid is vaporised, heat is required (i) to raise its temperature to boiling point, called the sensible heat, and (ii) to convert the liquid to gas at the same temperature, called the latent heat.

If the liquid is kept in a vessel at a high pressure, the boiling point is increased by the pressure; if the high-pressure liquid is allowed to pass into a vessel at a much lower pressure or a vacuum, it immediately vaporises and this abstracts heat from its surroundings.

The same principle applies to both types of refrigerators, but in the compression type freon is used, while the absorption type uses ammonia as the refrigerant, with hydrogen as the inert gas in the evaporator. In either method the evaporator was originally maintained some 11°C below the temperature required in the refrigerator but the most modern refrigerators have larger and colder freezing compartments (four star) as well as being provided with automatic defrosting facilities; the freezer requiring a temperature of −18°C or lower, and the auto-defrost type have a varying temperature differential over the on-and-off cycle from −20°C to several degrees above freezing point in order to achieve periodical defrosting. Heat passes from the air and food in the cabinet to the evaporator coils, causing any liquid refrigerant in the tubes to vaporise and change to gas.

Operation of compression refrigeration

A simple diagram is given in Fig. 114, in which the evaporator coil is inside the cabinet while the condenser coils are outside, either below or at the back, adjacent to the motor-driven compressor. The evaporator coil is around the ice-making compartment trays at the top, and the cold air falls to the lower spaces and is replaced by warmer air, from which the heat is abstracted. A single-cylinder, single-acting compressor is shown; the suction valve opens slightly before the piston reaches the bottom of its stroke and gas enters the cylinder. When the piston begins to rise both valves close and the gas is compressed; at a certain pressure, the delivery valve opens and the high-pressure gas passes to the condenser. A slight loss of heat liquefies the gas, the cooling process being assisted by cool air

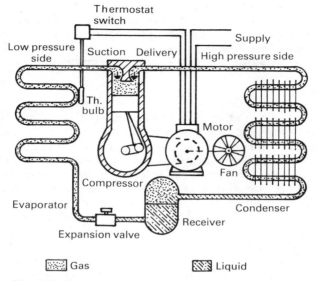

Gas   Liquid

Fig. 114 Principle of air-cooled compression refrigerator

from the fan blowing across the cooling fins. The final cooling is carried out at the bottom of the condenser, which contains high-pressure liquid. The compressor cylinder is kept free of liquid and only deals with the gas. The gas volume is very much greater than an equal weight of liquid, above which is space for some high-pressure gas. The receiver is connected to an expansion valve, which allows some liquid to pass to the evaporator; with loss of pressure and absorption of heat it changes back into gas. The cycle of operation is then repeated. A modern domestic compressor refrigerator with separate 4-star freezer compartment at the top is shown in Fig. 113.

Temperature control
A spirit-filled, sensitive, phial-operated thermostat controls the interior temperature within an adjustable range. The thermostat bulb is against the evaporator, and a capillary tube goes to the flexible copper bellows which operates the thermostat switch to start and stop the motor. The refrigerating unit's capacity is more than ample for producing the necessary refrigeration, so the motor is automatically on intermittent duty. With the absorption type, the refrigeration is more nearly continuous and there is not the same

reserve of capacity. In modern refrigerators with automatic defrosting the controls are a little more complicated so as to obtain a slower than normal change of temperature through a larger than normal range during the defrosting period, but the basic principle is the same.

The motor

Frequent starting and stopping, and ample starting torque for the compressor, make the capacitor motor most suitable, as it gives good starting torque, is silent in operation and free from radio interference. For d.c. supplies a compound motor is used. Domestic refrigerators need a motor output of from 120 W to 200 W depending on the cubic capacity of the refrigerator, which ranges from about 0·03 to 0·34 m³ in the usual domestic sizes. Small models use about 350 units, and the 0·17 m³ size about 500 units per annum. The electric refrigerator is an example of precision engineering and combines both electrical and refrigeration technologies so in the unlikely event of failure expert attention should be obtained.

Modern refrigerators use more advanced types of compressors than the one illustrated and some have the motor immersed in the refrigerant liquid but the principle of refrigeration is the same.

The freezing cabinet

This modern development of the refrigerator has gained popularity in the home for storing large quantities of perishable food for longer periods at lower temperatures than are obtained in the ordinary domestic refrigerator. It operates in the same way, however, and incorporates a larger capacity machine with more power for producing sub-zero temperatures. The electricity consumption depends on the size of the freezer, but it can be taken to be 1½ to 2 units per cubic foot per week. For the sake of economy it is important not to let heat from the room into the cabinet by leaving the door of the upright type open for longer than necessary as the warm room air and the cold air in the freezer refrigerator can change places very rapidly; but this is not such a serious matter with the chest type freezer having a top lid because the heavier cold air in the chest cannot move out to let the lighter warm air in. A typical freezer cabinet is included in Fig. 115 (page 197).

An important point is the siting of a freezer cabinet as it is a large piece of equipment and there is often insufficient space for it in the kitchen. The ideal location in such a case is in an outhouse, cool and dry in winter and summer, but definitely not the garage. It is dangerous to place any electrical appliance at low level in an

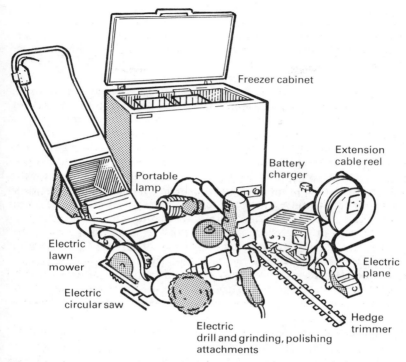

Freezer cabinet

Portable lamp

Battery charger

Extension cable reel

Electric lawn mower

Electric plane

Electric circular saw

Electric drill and grinding, polishing attachments

Hedge trimmer

Fig. 115   Some electrical appliances to be found in the modern home garage or garden shed
(See page 196–7 for warning regarding freezers in garages)

enclosed garage, as petrol, if leaking or spilt in an unventilated space, will vaporise and, being heavier than air, will spread at floor level and an electrical spark could then ignite the vapour with disastrous results. In addition, the average garage may be a damp location where severe condensation can form on the cold freezer surfaces causing electrical leakage, breakdown or corrosion of the steel outer casing. Freezers should also not be placed in glazed conservatories or other similar locations subject to high summer temperatures as excessive operation will result, causing a marked increase in electricity consumption. An average 10 to 15 cubic foot capacity freezer will consume about 1000 units per annum which at, say, 6p per unit, would cost £60 per freezer per year, so any additional cost would be very wasteful.

Home computers

The science of electronics has converted the mechanical calculator into the modern electronic computer for offices and has now extended its use in the form of the home computer. The introduction of the computer to schools was at first looked upon with some reservations that mental agility with figures would suffer, but it is now well established as a teaching aid and a skill in its own right and has come into the home for domestic use. The computer is so attractive and useful that it has been made available to all in the shape of small units for the pocket and larger equipments for industry but miniaturised so as not to present any space problems.

The home computer usually just plugs into the nearest socket outlet with a 3 A fuse in the plug top and its internal power pack provides all the different potentials that the circuits need to perform their calculations. The one main thing that usually causes a 'glitch' or 'hiccup' in the system is a stray voltage spike on the mains or induced on the earth connection by other electronic equipment on the same 415 V system or by switching surges and static discharges. The small pocket or desk-top computers do not suffer from this problem as they often have dry batteries for their energy supply or can even operate from daylight through photo-voltaic 'solar' cells.

In a closely packed group of houses, every 415/250 V circuit is likely to be connected to every other one (electronically speaking) and some filtration of the supply to each computer is the only answer to voltage 'spikes'. Special socket outlets can be obtained, interchangeable with the standard outlet, but with a mains filter circuit inside the socket outlet to clear away all mains borne interference. It is also a good idea to fit a separate 'clean' earth connection from the earth pin of the socket outlet for the computer and to run this in 4 mm p.v.c. cable physically separated as far as practical from any other wiring to an earth electrode outside the building. In extreme cases, it may be necessary to screen all mains and earthing cables to avoid radiated interference, or, failing this, move the computer to another floor on the other side of the house.

The basic principle of operation of a computer is in the behaviour of the electrons in a semi-conductor of electricity when submitted to varying stresses (from the applied voltages) according to the coded signals it receives and retains by the arrangement of circuitry and switching. (See page 3: *Effects of an Electric Current*). Hence, the first thing to be done in order to use a computer must be the preparation of a program designed to control the computer operation so that it indicates the correct results when 'given' a problem by the user. This is done by a 'programmer' who is versed in the 'language' in which the computer is intended to work and who produces a program 'record' which can be 'loaded' into the computer and to which the electronic components and circuitry in the computer will conform automatically; thereafter, the user operates the keys of the computer according to the code of the program stored in the computer in order to produce the result the user requires. It can be seen therefore that a computer is not a universal store of knowledge but simply a sorting or calculation device and a source of information it has previously been given.

Various ways of presentation of the results of computer operations are available such as pictorially on a screen or by the use of printers, and programs or data can be recorded onto magnetic tape or discs for storage after the computer is switched off.

Entertainment and education are also available with the home computer by the use of games and learning texts pre-recorded on tape or disc and operated, if appropriate, by control sticks or buttons plugged into the computer instead of the normal keyboard. Computer games are available in most large toyshops and educational programs are available from stationers and by post from specialist software (program) writers and distributors.

Garage and garden equipment

All garage socket outlets should be of metal-clad type mounted above bench level. Earthing is very important where used for small portable tools such as drills and grinders. All-insulated or double-insulated portable tools are becoming increasingly available; they are quite safe and avoid the necessity of earthing, but care is necessary to keep them in good condition and to avoid damage to the outer insulating shell, which could allow dirt or water to penetrate cracked plastic enclosures and make unearthed and exposed metal parts dangerous in the event of a fault in the equipment. But because of the mechanical strains that workshop and gardening appliances have to suffer they are still partially of metal construction, so efficient earthing is of paramount importance. Some makers incorporate a pilot lamp in the appliance which glows if the appliance is properly earthed and assures the user that the appliance can be used safely. Portable handlamps should be of the all-insulated type, in which it is impossible to touch any metal part of the lamp cap. Transformer equipment is available to provide low-voltage lighting for handlamps which makes portable apparatus absolutely safe to use, and the transformer steps the voltage down from 240 to 12 V. A portable handlamp with a strong wire guard and shade is useful in the house and garage and avoids the fire risk of using candles in dark places such as roof spaces and understair cupboards. A gripping attachment is useful as it leaves both hands free for working. Flimsy flexible should not be used; either t.r.s. (tough-rubber-sheathed) or heavy p.v.c. sheathed flexible should be used for tools; and for the handlamp a similar but lighter quality flexible should be fitted and examined periodically to see that it is in good condition.

Some of the electrical appliances usually kept in a garage or shed are shown in Fig. 115, but hand-held appliances should be kept on a shelf or in a cupboard for safety, not on the floor. The most common abuse that this equipment has to contend with is overloading. Most users are hobbyists, amateur gardeners or handymen who may not have had adequate experience in the handling of power tools, and so are liable to force the electric drill with its attachments to work harder than its design capability. Drilling metals and sawing wood with a circular saw are typical examples of this abuse. Overloads of this kind should cause the circuit protective device to open or to blow the fuse in the plug; and in some drills the maker has incorporated an automatic cutout, often a thermal switch, which operates when there is excessive heating due to overload and can be reset on cooling; this is a most useful design feature as it warns the user of overloading and promptly avoids damage to the drill or other attachment.

Drilling is always liable to break drills gripped in the drill chuck if the drill is not kept in perfect alignment with the hole being drilled, therefore it is advisable to have a 'sensitive' drill stand fixed to the bench for holding the electric drill when drilling objects or component parts are being made.

The best local protection for such tools or a lathe motor is the miniature circuit breaker or, ideally, an earth-leakage circuit-breaker with over-current tripping combined, in addition to the socket outlet plug fuse which might not be of the correct rating. A common type of electric drill now available has self-contained two- or three-speed gear change which enables choice of speed to suit large or small drill sizes. The hammer-drill is a useful variation and is best for drilling wall holes for plug fixings as it can make holes in a fraction of the time taken by a hand drill; special carbide-tipped drills are used for this purpose.

Electric gardening appliances are heavy-duty tools which require careful handling and earthing; they are also liable to rough usage. Where these tools are attachments for the electric drill it is necessary to fit them properly and firmly. Otherwise vibration may soon loosen the fixing and a loose attachment soon spoils its fit onto the drill. The body of the drill is generally of aluminium – a soft metal which wears rapidly if not gripped very firmly by the attachment – and before long the drill becomes useless for the purpose, for there is no simple remedy available to make attachments fit the drill again.

One of the problems in using electric gardening appliances is insufficient length of lead to reach the furthermost parts of the garden; the solution is to have an extension lead to add to the appliance lead. A good tough-rubber-enclosed coupler for the leads, with means of preventing inadvertent uncoupling, should be used to connect the leads together; and it is most important that the 'live' or supply side of the coupler has the sockets and not the pins – these should be on the appliance side. The handling of long flexible leads is also important (see page 180). Extension leads can be obtained on a special winding reel; these are particularly desirable because they avoid the trouble of coiling long flexible leads by hand.

Attention must be paid to Regulation requirements in respect of socket outlets for appliances to be used outside the house, which are given in Chapter 7 (page 113).

The caravan

The use of the trailer caravan or motor caravan for holidays is becoming increasingly popular and most of the better equipped camping sites provide electricity supplies for caravans from the public mains. While bottled gas will always be a standby energy supply for caravans when sited in remote places, the convenience of electricity, where available, makes a more trouble-free and safer source – a great consideration when on holiday. A 12 V installation connected to the car battery is, of course, a simple and fairly satisfactory scheme for lighting but it is quite inadequate for heating or other heavy loadings and even lighting can be an excessive drain on the car battery.

Many electrical appliances used in the house are desirable in the caravan; the shaver, fan, kettle, etc., especially as many modern caravans are provided with wiring and equipment for mains voltage supply.

Electrical safety is of paramount importance if a mains supply is used and Regulations have been drawn up to ensure this. This aspect is not less important than in the case of a house because the caravan is usually more combustible than a house and presents greater fire risks; and other mechanical considerations also present problems. Nevertheless, a safe installation can be ensured if the work is carried out properly and in compliance with Regulation requirements. In domestic (residential) fixed caravans the caravan installation must comply with I.E.E. Wiring Regulations in the same way as a house installation must comply.

Where a mains supply is offered to a mobile caravan, the site owner must provide automatic protection against an earthing fault by means of an earth-leakage circuit-breaker, a protective earthing conductor and a splash-proof socket outlet conforming to B.S. 4343, with two-pole and earthing contacts (in 6 o'clock position), located in a position within 20 m of the place intended for a caravan; this should normally have a capacity of 16 A unless a particular caravan needs a higher rating.

The caravan inlet connector must also conform to B.S. 4343 with two-pole and earthing contacts in the same way, and the connector must be located in a suitable recess with a lid on the outside of the caravan. The operating voltage must also be marked on the outside of this inlet. The important thing with mobile caravan connections is that the supply must terminate in a plug with socket tubes at the end of the flexible cable and the caravan inlet connector must have plug pins. This may appear to be opposite to within the house where an appliance has a plug with exposed pins to connect it to a socket outlet on the wall, but the reason is to ensure that unenclosed pins which could be touched are never live and that only concealed socket tubes are energised if the plug and socket is disconnected. The connectors must, of course, be non-reversible and have earthing contacts, with untouchable socket tubes at the supply point or supply end of the flexible lead and recessed contact pins on the caravan connector. In addition, a durable warning notice must be fixed near the main switch inside the mobile caravan, worded as follows:

INSTRUCTIONS FOR ELECTRICITY SUPPLY

To connect
1. Before connecting the caravan installation to the mains supply, check that:
    (a) the supply available at the pitch supply point is suitable for the installation in the caravan and its appliances.
    (b) the caravan main switch is in the OFF position.
2. Remove or raise any cover from the electricity inlet provided on the caravan, and insert the connecter of the supply flexible cable.

3. Remove or raise any cover from the socket outlet provided at the pitch supply point and insert the plug at the other end of the supply flexible cable.
4. Switch on at the caravan main switch.
5. Check the operation of circuit breakers, if any, fitted in the caravan.

IN CASE OF DOUBT CONSULT THE CARAVAN PARK OPERATOR OR HIS AGENT.

To disconnect
6. Reverse the procedure described in Paragraphs 2 to 4 above.

Periodically
7. Preferably not less than once a year, the caravan electrical installation should be inspected and tested and a report on its condition obtained as prescribed in the Regulations for Electrical Installations published by the Institution of Electrical Engineers.

The mobile caravan installation must be provided with automatic disconnection for operation in the event of a fault and should include a protective earthing conductor; internal wiring should be rubber- or p.v.c.-sheathed (or equivalent sheathing) and it must be effectively supported at intervals of not less than 150 mm horizontally and 250 mm vertically. These figures are increased to 250 mm and 400 mm respectively where the runs are in inaccessible places. In all other respects the I.E.E. Regulations apply as for a house installation.

Luminaires should be fixed directly to the caravan structure or lining. Pendants are unsuitable for caravans, but if used they should be provided with means of securing the lampshade and flexible cord against damage when the 'van is moved. Adequate ventilation of enclosed luminaires with tungsten filament lamps is important, especially between the fittings and the caravan structure. The best and most economical form of electric light for caravans is the miniature fluorescent tube; one 460 mm 15 W standard tube in a confined space, as in a caravan (which is comparable with 'local' lighting in a room) gives sufficient illumination for most recreational purposes. Smaller tubes still, of adequate wattage, are available for a 12 V d.c. supply from the car battery.

Earthing provisions must include bonding the earthing pin of the caravan inlet with the protective earthing conductor of the wiring,

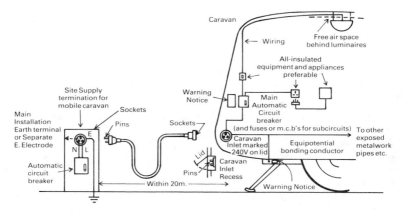

Fig. 116   Main safety requirements for mobile caravan

extraneous metalwork liable to become live, and the main structural metalwork of the caravan where a warning notice against removal must be fixed; the main equipotential bonding conductor must be not less than 4 mm² section; but this bonding does not apply to isolated metal parts if the caravan is constructed of insulating material.

Despite these provisions for earthing, the use of all-insulated appliances is preferable as an added safety measure. Where a caravan contains a fixed bath or shower compartment, the precautions previously described on page 106 under the sub-heading *Earthing and equipotential bonding* apply. In addition to the warning notice already quoted in this section, it is advisable that the testing and inspection of mobile caravans should be carried out at not more than three-year intervals or before reconnecting the caravan installation to a supply if it has been out of use for more than three months.

Most of the requirements described in this section are illustrated in Fig. 116.

11

Layout of a Typical Installation

The first thing to do is to mark a plan of the house with symbols of the electrical equipment required and then to prepare a corresponding schedule. If a plan of the house is not available, a sketch plan should be drawn approximately to scale.

Symbols for plans

British Standard Graphical Symbols are given in British Standard No. 3939, and an extract covering symbols commonly used in interior wiring installations is shown in Fig. 117 (page 206). Some of these symbols are used on the plan of the house illustrated later.

It will often be found that architects and installation engineers employ their own symbols, especially on small-scale plans where the standard symbols cannot be clearly marked, but such deviations should be minimal.

Installation materials

When the installation has been planned, the next consideration is the standard of installation to adopt. The use of steel conduit for wiring in houses as the highest standard has long been regarded as best practice; even where cheapness was important, steel conduit has been used with close-joint and split-type or grip-joint conduit fittings. With the development of plastics, however, plastic conduit has become much more suitable for rewireable house installations, and steel is now regarded as more appropriate for factory installations, public and other large buildings, and places where the strength and rigidity of steel is required. Conduits have

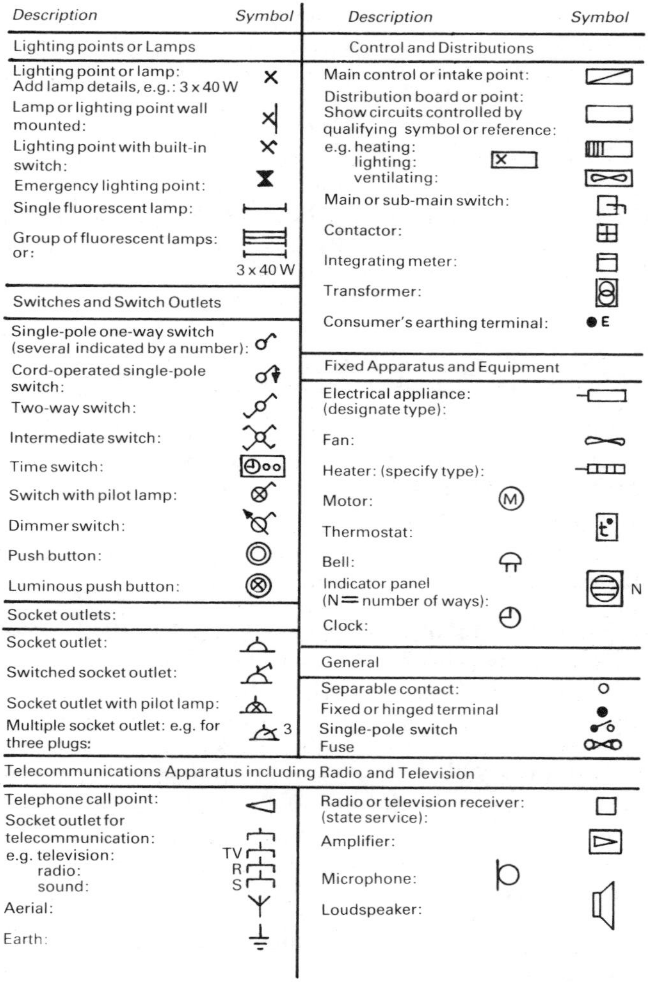

Description	Symbol	Description	Symbol
Lighting points or Lamps		Control and Distributions	
Lighting point or lamp: Add lamp details, e.g.: 3 x 40 W	✕	Main control or intake point:	
Lamp or lighting point wall mounted:		Distribution board or point: Show circuits controlled by qualifying symbol or reference:	
Lighting point with built-in switch:	✕	e.g. heating: lighting: ventilating:	✕
Emergency lighting point:	✗	Main or sub-main switch:	
Single fluorescent lamp:	⊢——⊣		
Group of fluorescent lamps: or:		Contactor:	⊞
	3 x 40 W	Integrating meter:	
		Transformer:	
Switches and Switch Outlets		Consumer's earthing terminal:	● E
Single-pole one-way switch (several indicated by a number):	ϭ	Fixed Apparatus and Equipment	
Cord-operated single-pole switch:		Electrical appliance: (designate type):	
Two-way switch:		Fan:	
Intermediate switch:		Heater: (specify type):	
Time switch:		Motor:	Ⓜ
Switch with pilot lamp:		Thermostat:	
Dimmer switch:		Bell:	
Push button:	◎	Indicator panel (N = number of ways):	N
Luminous push button:	⊗	Clock:	
Socket outlets:		General	
Socket outlet:		Separable contact:	○
Switched socket outlet:		Fixed or hinged terminal	●
Socket outlet with pilot lamp:		Single-pole switch	
Multiple socket outlet: e.g. for three plugs:	3	Fuse	
Telecommunications Apparatus including Radio and Television			
Telephone call point:	◁	Radio or television receiver: (state service):	▢
Socket outlet for telecommunication: e.g. television: radio: sound:	TV R S	Amplifier:	▷
		Microphone:	
Aerial:		Loudspeaker:	
Earth:			

Fig. 117   A selection of standard symbols for marking plans

the great advantages of long life, ease of wiring alterations or complete rewiring and cable protection; but, for lowest cost and the simplest form of wiring, the plastic sheathed cable system is very largely adopted nowadays in buildings with wood-joist floors.

The use of plastics is also adopted in the design of domestic switchgear and other wiring accessories. Switchgear for most house installations is made with an all-insulated plastic enclosure, and this type of equipment is most appropriate for sheathed wiring or plastic conduit systems, although metal-clad switchgear and fuseboards are often used instead, especially where larger sizes and capacities are not available in the all insulated types.

Plastics have also largely superseded rubber for cable insulation, and standard p.v.c.-insulated cables are made for up to 600 V to earth and 1000 V between conductors to B.S. 6004, and with butyl rubber insulation to B.S. 6007; p.v.c. and vulcanised rubber are used for most of the domestic flexible cords.

Whereas steel conduits are used as the earth-continuity conductor, plastic conduits necessitate a separate protective earthing conductor in all the conduits, together with the circuit wiring; with plastic-sheathed wiring, this must also have a separate protective earthing conductor within the sheath.

Layout considerations

The lighting in a new house should be considered in relation to the fixed furniture in the various rooms. The switches should not be behind doors or in inconvenient positions just to save a few feet of wiring. The position of lighting points and switches should be carefully selected before the wiring is begun so as to avoid later alterations with buried wiring; such alterations, involving replastering, often cause unsightly discoloration of the wallpaper. The occupier should be consulted if possible, but experience and common sense should be used in offering advice.

The service cable will be arranged by the supply undertaking, and their fuses and meters should be adjacent to the point of entry but should not take up valuable cupboard or pantry space. With some modern houses the garage is often convenient for this, or a special cupboard or inset recess in a wall can be provided which will also house the main distribution boards for lighting, power and other purposes. It is a growing practice to arrange such a position with a small window or covered opening to the outside or a metal box with door on an outside wall for the meter and service equipment so that the meter-reader can see the meter without having to enter the house.

We now look at the installation for a typical detached house in which no solid fuel or gas is used and electricity serves all purposes. The loadings are arbitrary, as the user would no doubt choose lamp sizes or the luminaires to be installed to suit himself, and these details are not usually known at the design stage.

Electrical installation considerations

After preparing a layout plan a schedule of lighting and sockets is made, which also shows other information and lighting data for the various rooms. The sizes of socket outlets have been confined to 13 A, and these would be wired on ring circuits. The use of twin 13 A socket outlets for all positions in the kitchen, main rooms and bedrooms would be very advantageous. The electric clock points, shaver sockets and the bell transformer take only a small current, so they have not been scheduled. The positions of switches and various outlets are indicated on the plan (Fig. 118, page 210) by standard symbols, but the conduit runs are not shown. It is a good idea to arrange the larger rooms so that they are not dependent on one sub-circuit; and, if the house load necessitates more than a single phase of the supply to be brought in with load balancing on two or three phases, it is important to avoid outlets on different phases being on the same floor, and certainly not in the same room, as line voltage (415 V) would exist between conductors possibly within reach and therefore dangerous.

Schedule of lighting and sockets

	Size m	Area m²	Lamps	Watts/ room	13 A sockets
Sitting room	$5 \cdot 0 \times 3 \cdot 5$	17·5	3×60 W(T)	180	6
Dining room	$3 \cdot 7 \times 4 \cdot 1$	15·2	1×150 W(T)	150	6
Study	$2 \cdot 6 \times 3 \cdot 0$	7·8	1×100 W(T)	100	4
Hall	$2 \cdot 4 \times 3 \cdot 5$	8·4	1×60 W(T)	60	1
Lobby, lavatory, W.C.			3×40 W(T)	120	
Kitchen & cupboards	$3 \cdot 8 \times 2 \cdot 4$	9·1	2×40 W(Fl) 2×40 W(T)	180	4
Laundry room	$1 \cdot 8 \times 1 \cdot 7$	3·1	1×60 W(T)	60	1
Tool shed	$1 \cdot 8 \times 1 \cdot 2$	2·2	1×40 W(T)	40	
Porches front and side			2×40 W(T)	80	
Garage	$2 \cdot 6 \times 4 \cdot 5$	11·7	2×60 W(T)	120	2
Ground floor total		75·0		1090	24

	Size m	Area m²	Lamps	Watts/ room	13 A sockets
Bedroom 1	5·0 × 3·5	17·5	1–60 W(T) 3–40 W(T)	180	6
Bedroom 2	3·8 × 4·2	16·0	1–60 W(T) 2–40 W(T)	140	6
Bedroom 3	3·8 × 2·5	9·5	1–60 W(T) 2–40 W(T)	140	4
Utility room	2·4 × 3·0	7·2	1–40 W(Fl)	50	2
Bathroom	2·5 × 2·1	5·2	1–60 W(T)	60	
W.C.			1–40 W(T)	40	
Top of stairs landing			2–40 W(T)	80	1
Airing cupboard			1–40 W(T)	40	1
First floor total		55·4		730	20
Total for house		130·4		1820	44

Details of room sizes, lighting and socket outlets

The supply authority's fuses and meter are shown in the garage at A, mounted at a reasonable height on the wall and adjacent to the incoming service cable. The consumer's main switches and fuses can also be at A. An alternative point of entry is shown to the left of the front door, with the distribution fuseboards at B at the back of the hall but this could be with the main switchgear or combined in a consumer unit if desired. The final choice of positions depends on local requirements. In either of these alternatives, it would be possible to arrange the meters behind a glass panel or small window in the wall to enable the meter-reader to inspect without having to enter the house and a metal or plastic meter cabinet is fitted in or on an outer wall for this purpose where appropriate and acceptable.

Wiring sizes and loading
The total general lighting load is 1820 W, which with a 240 V supply gives a current of 7·6 A, though all this will never be on at one time. Although 2 A socket outlets could be used for lighting purposes

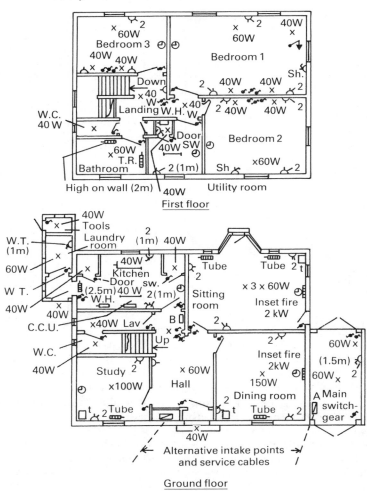

Fig. 118   Plans of a detached house

instead of some of the 13 A sockets shown and be wired to the lighting circuits, the saving would be small and the usefulness of more numerous 13 A sockets for all purposes would be lost.

Considering the ground floor, the maximum lighting load is 1090 W, which requires 4·5 A at 240 V. As there are seventeen points, two circuits should be run from the distribution boards with 1 mm^2 cables.

Reference to Table 7 in Chapter 6 (page 85) will show that voltage drop calculations may be ignored, as the maximum length of run of the circuits is so short. There are twenty-four 13 A sockets, and these may be connected to a single 30 A ring circuit with 2·5 mm^2 rubber- or p.v.c.-insulated cable, or 1·5 mm^2 mineral-insulated cable.

On the first floor the maximum lamp load is 730 W, which requires 3 A at 240 V. There are sixteen points, and again two circuits are desirable to avoid the whole floor being in darkness if a fuse 'blows'. Regulations do not limit the number of lighting points on a circuit but the circuit current is limited to 15 A load, with each lamp rated at 100 W. Thus the smallest p.v.c. cable of 1 mm^2 section in a group of three with a capacity of 12 A could feed up to twenty-nine lamps. There are twenty 13 A sockets, which may be connected to a single 30 A ring circuit.

As this will be an all-electric house, we will assume adoption of the standard domestic tariff, and there will be no need to keep lighting, heating and power points in separate installations for separate metering. With a single-phase, two-wire service, there-fore, four lighting circuits, two ring circuits for socket outlets and separate circuits for an electric cooker and water heater, one distribution fuseboard with six 15 A ways (one spare and four fused 5 A) and three 30 A ways will be adequate for the whole house and, if expense will permit, a miniature circuit-breaker distribution board will provide a high standard of installation.

Details of lighting, switching and socket outlets
The sitting room is shown with a central luminaire for general lighting, containing three 60 W lamps controlled by two switches for using one, two or three lights. The two-way switches by inside and outside doors control the single lamp. The switches are mounted near the doors, at the same side as the handles, 1·3 m above the floor. A common height should be maintained throughout the house. 13 A socket outlets are provided on each wall between the doors and the fireplace to avoid flexible cords crossing them, and there is a built-in electric fire which can be connected to the ring circuit through a fused spur unit and a separate double-pole switch if the switch is not incorporated in the electric fire itself. The other fixed heaters can be connected by socket outlets, or in the same way as the fixed fire, but with pilot lamps where necessary. The sockets should be fixed above the skirting board to give room for the flexible

cord to turn out from the bottom of the plug; 450 mm above floor level is ideal and avoids reaching to floor level to insert plugs or operate the switches of switched sockets. For use with bench or worktop equipment, the sockets should be 150 mm above bench level. A permanent fixed clock point is provided above the fireplace and is connected to a lighting circuit; this obviates trailing leads to the synchronous clock. Clock points are also indicated in other rooms.

The dining room has a 150 W central luminaire and the previous remarks about socket outlets also apply to this room. The use of extra standard-lamps is intended to boost the lighting locally as required in both these rooms.

The study is provided with a 100 W central luminaire, but if the walls are dark with many bookshelves a greater wattage would be needed; it is assumed, however, that a desk or table lamp would be used here to supplement the general light. The positions of the various outlets depend on the wishes of the occupier and the distribution of the furniture. No bracket lighting has been indicated, but decorative wall brackets are extremely popular in living and drawing rooms, and suitably located points for these should not be overlooked. 13 A sockets for the vacuum cleaner are provided in the ground-floor hall and first-floor landing.

The kitchen has two 40 W fluorescent lamps in enclosed luminaires, which, it is suggested, should be positioned to give good illumination to the cooker and sink on one side and to the work table between the larder and a cupboard which gives access to the serving hatch on the other. In each of these cupboards a 40 W lamp is mounted in a ceiling luminaire, controlled by door switches which go 'on' when the doors are opened. A similar arrangement is fitted to the airing cupboard on the first floor. The main lighting is controlled by two two-way switches at each door.

The laundry room has an enclosed luminaire, the switch being watertight; the porch light is an enclosed wall luminaire with the switch in the kitchen; and the hall and first-floor landing are provided with 60 W and 40 W lamps respectively, both with two-way and intermediate switching.

The garage is provided with a general light (which might be better placed at the end where the car bonnet will be) controlled by two two-way switches, a light over a bench and two 13 A sockets for a portable electric tool and handlamp.

The bedrooms are provided with a 60 W lamp over the dressing table, and 40 W lamps over the beds and in front of a wardrobe; alternatively, wall brackets or portable bed-table lamps can be provided and connected to twin or separate socket outlets by the beds. The switches are shown near the door, on the same side as the door handle, and two-way switching is provided from the bed and door positions. Ceiling switches can be used for the wardrobe light, and bed switch and wall lights over bedheads instead of ceiling points if desired.

The bathroom is provided with a 60 W enclosed ceiling luminaire controlled by a cord-operated ceiling switch. If the bathroom is to be provided with a washbasin and mirror, an all-insulated wall lighting point is desirable over the mirror, and possibly a shaver supply unit as well.

The illumination of the utility room should be generous, hence a 40 W fluorescent lamp is used with socket outlets for a portable lamp and the sewing machine or other equipment.

The other lighting arrangements do not call for any special comments, except that the laundry room and porch luminaires should be of watertight pattern, and that the sizes of lamps given can be increased with advantage if higher standards are desired.

Schedule of heating and other outlets

It is convenient to consider other fixed heating and power outlets separately, and make a schedule of these; the various outlets being indicated on the plan (Fig. 118, page 210). The 13 A socket outlets listed in the lighting schedule will provide for all the portable apparatus likely to be used, including portable fires and kitchen equipment, but there are fixed fires, tubular heaters and other non-portable apparatus which should be listed. With the exception of the cooker, these outlets are connected through switched fused spur units to the nearest ring circuit for 13 A sockets, as shown in Fig. 57, page 93, using a suitable size of cable for the apparatus to be connected but ensuring that the spur fuse is not too large to protect the cable and, in any case, not larger than 13 A rating.

In the dining room and the sitting room 2 kW built-in fires are fitted, while a socket outlet in the study can be used for a 2 kW portable fire, convector or fan heater if desired. In the three main ground-floor rooms, tubular or skirting heaters are fitted under the windows, each consisting of a bank of two tubes 1·2 m long taking 480 W, or equivalent, which on a 240 V supply requires 2 A, so 1 mm^2 cable is large enough. Thermostats are mounted in the corners of the rooms on the outer walls so as to switch these heaters off when the room temperature exceeds 15°C. When the other electric heating is not in use, these heaters keep the rooms aired in cold weather. A similar arrangement of heaters could be adopted for the bedrooms, but our scheme assumes the use of portable heaters in these rooms. A portable convector heater can be used in the hall. The most popular method of warming the bathroom is to use a wall-mounted reflector heater, as illustrated in Fig. 97(a) page 156, fitted 300 mm below the ceiling and efficiently earthed. An electric towel rail, 750 mm long and loaded at 75 W, is provided. These points can be connected to the ring circuit through switched spur units. All the exposed metalwork of electrical equipment, and the metalwork of other services also, must be electrically bonded together in the bathroom.

Schedule of heating and other outlets

Room	Heaters	Other apparatus
Sitting room	1 2 kW fire 2 tubular, 960 W	
Dining room	1 2 kW fire 1 tubular, 480 W	
Study	1 tubular, 480 W	
Kitchen	Wall heater, 750 W	Refrigerator, Cooker, 12 kW, Water heater, 1500 W,
Laundry room		Washing machine with heater 2500 W.
Bathroom	Towel rail, 75 W 1 wall heater, 750 W	
Linen cupboard		Water heater, 1500 W

The kitchen contains the cooker, which is fed from a separate fuse in the consumer unit and for which the load is ascertained by taking the sum of the first 10 A of the total cooker loading, 30% of the remainder and 5 A for the socket outlet in the control unit. This results in a capacity of less than 30 A for the control unit and circuit cable. This circuit is wired back to a separate fuseway in the consumer unit or, if necessary, to a switch fuse unit at the main switchboard. A pair of 4 mm² p.v.c. cables will carry 32 A. The 90-litre water heater, rated at 1500 W, takes 6·25 A and is connected to a switched spur unit with a pilot lamp fed from the nearest socket ring circuit, or to a separate fuseway in the consumer unit using 1 mm² cable, thus obtaining greater available capacity in the ring circuit.

If extra heating is required in the kitchen for cold mornings, a portable heater can be used, or a high-level heater could be fixed on the wall – as for the bathroom – and a suitable outlet point provided. The refrigerator is connected to a 13 A fused plug fitted with a 3 A fuse.

Separate water heaters for kitchen and bathroom would be more economical by saving heat losses in long runs of pipework if a single water heater was installed. The kitchen water heater also supplies the adjacent lavatory basin, and the bathroom water heater in the airing cupboard could also supply basins in the bedrooms if desired, but the design of an electric water-heating system is a special subject and is referred to in Chapter 9.

If a water-heating cylinder is not installed in the linen cupboard, a small tubular heater would be required here.

The laundry room is provided with a watertight 13 A socket outlet mounted well above the floor for the washing machine.

Estimation of current

All this load will not be on at once, and diversity of use can be allowed for in determining the capacity of the main cables and switchgear. This is done as follows, using the allowances for diversity given in the I.E.E. Wiring Regulations:

		Ampères
Lighting:	66% of 1820 W = 1214 W:	5
Cooker:	say, 12 kW; first 10 A:	10
	30% of remainder:	12
	If socket-outlet in control unit:	5
Water heaters: 100% of 3000 W:		12·5
Socket outlets and stationary appliances:		
	100% of largest fuse rating of individual circuits:	30
	40% sum of remaining fuse ratings:	12
	Total:	86.5A

Main supply and load balance

Having estimated the total electrical load in the house, the supply authority must be approached and asked whether this load can be connected to a single-phase, two-wire supply. If so, this figure can be rounded off to 100 A for the capacity of the main switch fuse or circuit-breaker, and the main cables must therefore not be less than 25 mm^2 in section for p.v.c.-insulated cables. In the event of the supply authority requiring the load to be balanced on two or three phases of the supply, the whole distribution system would have to be designed differently and divided into two or three independent sections with equal loads, as far as practicable, but so arranged that the wiring to all outlets in the same room are on one and the same phase. When possible this rule should be applied to a whole floor to ensure maximum safety and the avoidance of shock at medium voltage (415 V). Fig. 119 shows how such an installation would be arranged for balancing on a three-phase, four-wire supply if necessary.

Variations in domestic installation wiring

A development of the spur from ring circuits for other 13 A sockets is to feed the lighting circuits from spur boxes fused at 3 A, taking

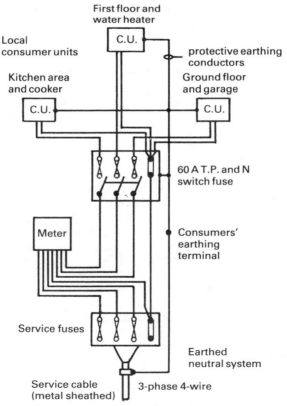

Fig. 119 Main supply and distribution switchgear for three-phase distribution in house

lighting wiring up through the switches to the luminaires on walls or ceilings, as shown in Fig. 120. This reduces the length and cost of lighting circuits, and the size of the consumer unit, but it is not advisable for the larger installation, where it is best to keep the lighting completely separate from the socket-outlet circuits. Water heaters and other fixed appliances may also be connected to spurs from the ring circuit. A typical small house ground floor is shown.

In some installations with the distribution point in a central position, it may well be more convenient and economical to use radial circuits for the 13 A socket outlets instead of the ring circuit.

A method of wiring more suited to large housing schemes than to individual or small numbers of dwellings is the harness system, whereby the wiring for a house or flat is designed and constructed in the factory, so that the wiring 'harness' is placed in position, and the

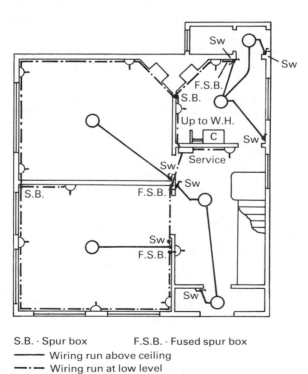

S.B. · Spur box   F.S.B. · Fused spur box
—————— Wiring run above ceiling
— · — Wiring run at low level
════════ Wiring to cooker and water heater

Fig. 120   Ring circuit with socket outlets and spurs to lights

outlets and consumer unit fixed and connected with the minimum of labour on site. This arrangement usually involves a central junction box placed in the floor or roof space, with radial circuits of sheathed wiring or cables in plastic small-bore tubes from this box to all the equipment and accessories of the installation.

The kitchen

The many different labour-saving devices found in a modern kitchen generally include an electric cooker, refrigerator, washing-machine, tumble drier or spin drier, dish-washer, waste disposal unit, food-mixer, percolator, electric kettle and toaster, therefore the layout of the kitchen should be such as to give the most complete service with ample and convenient socket outlets for connection. This cannot be achieved with only one or two socket-outlet points fitted at positions that are assumed to be suitable. An ideal arrangement would be a pair every metre around the kitchen, and as neat wall trunking fitted with 13 A sockets at any required spacing is

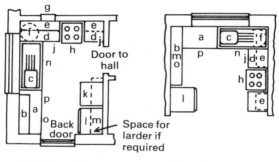

(a) L-shaped layout   (b) U-shaped layout

Fig. 121   Efficient planning of essential equipment in a small kitchen

Key: *a* Preparation counter
      b Ventilated food store over
      c Sink
      d Service and dishing-up counters
      e China and glass store over
      f Water heater under
      g Service hatch
      h Cooker
      j Pans and utensils under
      k Washing machine
      l Refrigerator
      m Grocery store
      n Garbage can under sink or waste disposal unit
      o Vegetable store under
      p Space for drier

(A dishwasher might stand over *f* or under draining board)

already used in science laboratories there is no reason why this scheme should not be adapted to domestic kitchens over worktops.

The layout of kitchen furniture and apparatus should be coordinated under the following headings:

1. Relative disposition of equipment for operations.
2. Basic requirement of apparatus, e.g. water, electricity.
3. Objective visibility under natural and artificial lighting.
4. Safety and convenience with minimum fatigue in use.

Classic layouts following the basic preparation-cooker-sink disposition of equipment with L-shaped and U-shaped planning are shown in Fig. 121, but lighting, socket outlets, electric clock, etc., are not shown for clearness. Most home and women's magazines have advertisements showing modern layouts with island or extension worktop cabinets fitted with built-in cooking equipment and controls where there is enough space for these and specialist contractors can be employed to design and provide furnishings and equipment for any size and shape of kitchen.

12

Radio and T.V., Aerials, Recorders, Telephone Communication and Static

Radio phenomena are very different from ordinary electrical phenomena, such as those explained in Chapters 1 and 2, and require an entirely different approach. We have learnt that an unchanging direct current produces a steady magnetic field, but that no effects are induced in other conductors in the vicinity until the current is switched off and drops to zero; this sudden change does, however, produce an inductive effect both in the circuit itself and in adjacent conductors, particularly if the circuit is an inductive one, i.e. a coil with a magnetic field. Alternating current, which is continuously varying, therefore always involves inductive effects, whether in a single straight conductor or in a coil. Inductance depends on the number of linkages of the current flow with a magnetic field; it is represented by L and is measured in Henries (symbol H). It results in a self-induced voltage, diametrically opposed to the applied voltage and lagging a quarter cycle (90°) behind the current cycle (as shown in Fig. 122 (a)) and a reactance, X_L, which is expressed numerically by $2\pi f L$ ohms (f = frequency, Hz).

Similar considerations apply to the stressing of an insulating medium between conducting surfaces, i.e. the capacity effect. No current flows in a condenser or a capacitor with an unvarying potential applied to it, but with an alternating potential the insulating medium is stressed alternately in both directions according to the frequency and a current flow is effected. The capacitance is measured in farads (symbol F) and is represented by C. It relates the charge of electricity to the voltage it produces. The numerical considerations are exactly as for inductance effects, but the self-induced voltage is 90° in front of or leads the current cycle (as shown

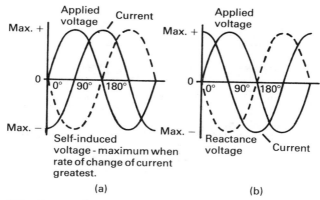

Fig. 122 Current displacement with inductance and capacitance
  (a) Inductance only
  (b) Capacitance only

in Fig. 122(b)), so its effect is directly opposite to inductance and its reactance, Xc, is expressed by

$$\frac{1}{2\pi fC} \text{ ohms.}$$

This theory and formulae are used in many a.c. calculations with mains voltage and frequency (50 Hz) but with the very much higher frequencies of radio waves it can be seen that values of L and C become much more important in the control and tuning of radio circuits to match the numerous wavelengths involved.

With such very small radio-frequency currents, the resistance of aerial and feeder conductors is of very little account and is neglected, the main factors being inductance and capacitance with their resonating effects – analogous to the tuning fork and vibrating strings in musical instruments; the relation between these two factors is expressed by the impedance value,

$$Z = \sqrt{\frac{L}{C}} \text{ ohms, as a measure of resonance.}$$

It will be seen that both inductive and capacity effects are functions of the alternating frequency. At low values, such as mains supply frequency, the effects are limited to the local circuits, but as the frequency is increased these effects have much greater electro-magnetic effects on the medium known as the ether, which we

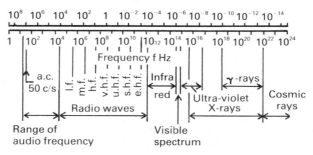

Fig. 123　Electromagnetic wave spectrum

believe permeates all space and matter but is most responsive in space and the atmosphere. The ether has a high degree of elasticity and response to high frequencies, especially those above 1000 Hz; in free space the oscillations or waves will travel unimpeded for millions of miles, but on earth they are attenuated or absorbed in varying degrees by the atmosphere and other objects in their vicinity. The electromagnetic wave spectrum is shown in Fig. 123, and with our ready knowledge of the behaviour of light rays in the visible part of the spectrum we can begin to understand to some extent how the other kinds of waves at different frequencies operate. Since all radiation waves travel at the speed of light, which is 300,000,000 metres per second, the wavelength and frequency can be related by dividing the velocity by the frequency, i.e. the wavelength (symbol λ, the Greek letter *lambda*)

$$= \frac{300,000,000}{\text{Hz}},$$

or, since we are dealing with high values of frequency,

$$\frac{300,000}{\text{kHz}} \text{ or } \frac{300}{\text{MHz}}.$$

Having arrived at some understanding of what radio waves are, we can proceed to discuss the method of producing and receiving them for broadcasting purposes. The aerial is the instrument used for stressing the ether and making it oscillate at the desired frequency to carry the signal it is required to broadcast, and for receiving the signals. Transmitting and receiving aerials are exactly the same in principle and operation, and any aerial can be used for both purposes.

Radio and television aerials

By imposing a suitable voltage on an aerial system, radio waves are generated in the ether with a wavelength depending on the frequency of the input and the dimensions of the aerial. The aerial is designed to be in 'tune' or to resonate with the applied frequency by making the length of the aerial conductor correspond to the length of the half-wave so as to be able to carry each full half-wave of current from zero to maximum and then to zero again over its length in alternate directions with the radio wave. The inductance of the aerial distributed over its length and its capacitance between the two halves are calculated to resonate or be in tune with the required frequency and wavelength. The mean of the Band wavelengths given in Table 19 is taken for this purpose. Thus maximum signal strength or gain and selectivity (of wavelength) is said to be obtained, and this is augmented by suitably directing the aerial towards the transmitting station.

The common basic aerial design is the half-wave dipole, which is formed of two metal rods or tubes in line, each a quarter wavelength long and connected at the inner ends to the two conductors of the

Table 19 Frequency and wavelength of broadcast bands in UK

Aerial groups	Channels	Frequency bands	Approximate wavelength, m	Term and use
Sound radio	—	kHz 150–285	2000–1185	Long wave sound (l.f.)
Sound radio	—	525–1605	560–188	Medium wave sound (m.f.)
Sound radio	—	MHz 2–30	150–10	Short wave sound (h.f.)
Band II radio (programmes 2, 3, 4) and stereo	—	87–100	3·4–3	f.m. sound (v.h.f.)
Band III	6 to 13	174–216	1·65–1·4	T.V. (v.h.f.)
Band IV television	21 to 34	471–582	0·64–0·52	T.V. (u.h.f.)
Band V	39 to 68	615–960	0·49–0·35	T.V. (u.h.f.)

(Band I was v.h.f. for black and white television which has now been phased out.)

downlead or feeder cable. The dipole aerial is fixed vertically for vertically polarised transmissions and horizontally, broadside on to the transmitting station direction, for horizontally polarised transmissions. The Band II radio signals are horizontally polarised, as are some broadcasts in other bands; others are vertical. Table 19 gives the frequency and wavelengths of broadcast bands used in the UK.

The list of broadcasting stations within each group is too long to include here, but aerials are made for each group or frequency band to give reasonably good results for any frequency within the group, and for best results a separate aerial should be provided for each group required. Groups of four channels are allocated to each main area station in the u.h.f. television bands for different programmes, the fourth being reserved for a future additional programme; and over fifty main stations with some three hundred relay stations will eventually transmit colour television to almost the whole of the UK population.

The television picture is made up of a large number of horizontal lines by which the cathode ray scans the screen from top to bottom, with such imperceptible speed that the eye accepts the picture as an instantaneous whole, much as it views a cinema screen. In Bands IV and V (v.h.f.) the 625 line system for the screen picture is used which has much better definition than the old 405 line system.

Vision and sound signals for the same broadcast are transmitted at slightly different frequencies, being only about 3 MHz apart; each station has about 8 MHz bandwidth per channel, and this has been worked out to enable reasonable selectivity and freedom from interference to be obtained from the standard aerials for each band. The frequency and wavelength for each transmitting station refers to the carrier wave on which the signal is imposed by variations of amplitude or frequency, hence the terms amplitude modulation (a.m.) and frequency modulation (f.m.). The power output of small broadcasting stations may vary from 5 to 50 kW and provides reasonably good reception within a radius of about 40 to 80 km, but the main stations have powers from 100 to 1000 kW and cover much larger areas, up to 160 km radius or more, according to the height and contours of the area.

Good reception of radio and television transmissions depends primarily on locality and distance from, as well as the output strength of, the transmitting station. Hills, valleys and large structures have a marked effect on the strength of signals received, and site tests of signal strength are often the only way of determining this before a decision can be made on the type of aerial installation to use in areas remote from stations. Tall structures obstruct or reflect signals according to their position in relation to the direct line between station and receiving aerial; this effect may involve the adjustment of aerial direction and is often the cause of 'ghost' or multiple images on the television screen, even in good reception areas.

In some urban areas television and radio relay services by cable are available. In this system a central receiving station has an efficient and well-sited aerial array and amplifying equipment, and redistributes the broadcasts over land lines to subscribers in the area – much in the same way as the public telephone service. The wires are taken into each subscriber's house and a programme selector is provided. Receivers may be hired or purchased. Individual aerials are therefore not necessary with radio relay services.

No one will dispute the ugliness of numerous aerials on the roofs of rows of houses, and in estates where this is not permitted, or in blocks of flats where it may be impracticable, central aerial systems and amplifying equipment are installed and cables distribute the signals to the tenants' houses, where their own receivers can be used in the same way as with an independent aerial.

Most houses, however, have independent aerials, and these are designed, firstly, to suit the waveband of the required transmission and, secondly, to suit the location and signal strength available. The shorter waves are more like light waves and rapidly weaken as they go out; they pass through the ionised layers of the upper atmosphere – whereas the long waves are reflected – and are reflected by hills and tall structures. So, the higher the aerial, the better the reception will be, although in areas within a few miles of the transmitting station good reception is often obtained with a suitable aerial mounted in the loft or roof space, rather than outside on a chimney stack, which makes for a better appearance of the house with cheaper fixings and less maintenance.

Aerial design

The basis of aerial design is to provide a conductor of almost half-wave length, as this resonates to a maximum with the carrier waves and is said to be in 'tune' with the wavelength. In practice, the length is a compromise since it must receive the different sound and vision wavelengths from the same station simultaneously, as well as signals from other stations in the same waveband. This is assisted by making the aerial conductor as large as possible in section to widen its response to waves within the required bandwidth, and for this purpose it should be about 15 mm diameter at least and is usually made from light alloy tube. Aerial makers generally add the channel number or a letter to their catalogue numbers for convenience to designate each kind of aerial; these are as follows:

> A   for channels 21 to 34 colour code red
> B   for channels 39 to 53 colour code yellow
> C/D for channels 48 to 68 colour code green
> E   for channels 39 to 68 colour code brown.

The basic dipole aerial can be fitted with what are known as parasitic elements to improve the gain and the directional properties of the aerial. These elements have no direct electrical connection to the aerial of any importance but serve to reflect or direct signal strength to the aerial proper, according to whether they are situated behind or in front of it relative to the signal direction. When behind as reflectors they are a little longer than the dipole and when in front a little shorter, the spacing being a quarter wavelength from the dipole. The effect is to make the dipole more sensitive in the direction of the required signal and less sensitive in other directions where interference may be located. Fig. 124 shows a dipole aerial (H-type) with reflector and director elements and the resulting

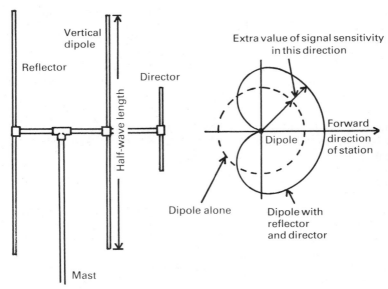

Fig. 124   Vertical dipole aerial with reflector, director and polar diagram (in horizontal plane)

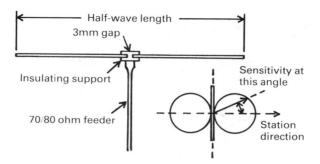

Fig. 125 Half-wave dipole horizontal aerial, showing how feeder cable is connected at centre, and polar diagram

polar diagram, which indicates the relative sensitivity at different angles around the dipole. It will be seen that, while the reflector increases the signal gain in the direction of the transmitting station, it cuts out entirely any unwanted signals or interference within a fairly wide angle in the opposite direction. A horizontal dipole is illustrated in Fig. 125, and the polar diagram shows that the response at right angles to the required signal is negligible over an appreciable angle. The gain is increased by adding director elements as shown in Fig. 126(c) (page 228). Unfortunately this lowers the impedance but this can be restored by folding the dipole as shown.

With the higher ranges of frequency, more parasitic elements can be used with advantage. There are combined aerials for receiving more than one band on the same aerial array and various shapes of aerial designed for special purposes, but they are seldom as efficient as the single-band aerial designed and used for its own purpose. Fig. 126 shows various outdoor aerials commonly in use at the present time, but it is not practical to show all the varieties and combinations obtainable, nor to deal with the theory of their design more fully in this book.

The main station transmitters use horizontal polarisation (the aerial elements are in the horizontal plane) with u.h.f., and the receiving aerials must be in the same plane. Local relay stations in areas of poor reception use vertical polarisation and the receiving aerial's elements must be arranged vertically.

Mention should be made of the frame or loop aerials with from three to eight turns of wire, spaced apart on 2·4 to 0·9 m square frames and having natural wavelengths of 160 to 185 m respectively.

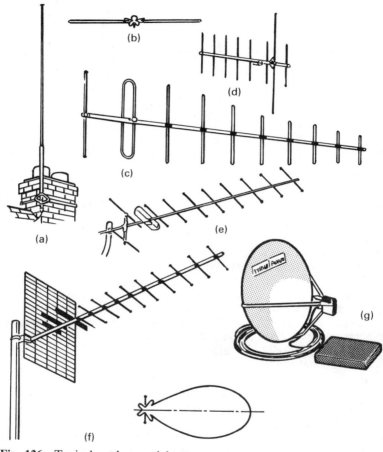

Fig. 126   Typical outdoor aerial arrays
   (a) 5·5 m skyrod aerial for long, medium and short wave recep-
        tion, lashed to chimney
   (b) Single-element dipole, horizontal, for Band II
   (c) Ten-element vertical array for Band III with folded dipole
        (centre), reflector and director elements
   (d) One + seven element array for Bands I and III combined
   (e) Ten-element array for Band IV, medium range (10–20
        miles)
   (f) Band V thirteen-element aerial with mesh reflector and
        polar diagram
   (g) Dish aerial for satellite transmissions

This aerial is directional with a figure-of-eight polar diagram, with a maximum response in line with the turns. Small frame aerials were once made for portable radio sets, but they have been largely superseded by the ferrite rod aerial for medium and long waves. The core material has a high magnetic permeability; it is responsive up to about 100 MHz, and very sensitive aerial coils on a ferrite core can be provided in a very small space for v.h.f. and u.h.f. signal reception. These aerials can be used for direction-finding, and this is evident from the directional qualities of the small portable transistor sets that use them. Loft and indoor aerials can be used where the signal strength is sufficient, and 7 to 10 m of insulated stranded copper wire hung in a roof space can be very effective for long and medium waves; but with limited space smaller aerials are also available, and these are designed to pick up a useful fraction of the wavelength. In areas of good signal strength, it is possible to use a simple wire as a f.m. dipole fixed on a suitably orientated picture rail in a room, with a piece of ordinary twin parallel (not twisted) flexible lead to the set for Band II f.m. reception. The overall horizontal length of the dipole should be 1.59 m, and the lead should be connected at the centre where the wire is cut into two halves.

Downleads

Where signal strength is good and interference small, a plain insulated downlead can be used for long, medium and short wave reception with aerials such as (a) in Fig. 126, but where interference is troublesome the use of special feeders or downleads and matching transformers may be necessary. For all the television bands it is necessary to use specially designed downlead cable in order to increase the aerial gain and minimise interference, which is more troublesome in the higher frequency bands.

The aerial and its feeder cable must be taken together in considering the natural wavelength or tuning resonance of the aerial installation. The old long-wire outdoor aerial of the early years of radio broadcasting, with a length, including the downlead, of about 30 m, would have a natural wave-length of about 120 m (the maximum permitted length is about 46 m). The receiver tuning device is then combined with the aerial and feeder to adjust the overall resonance to match exactly the wavelength of the station to be tuned in. The downlead can pick up interference that is missed by the aerial, such as radiation from the ignition systems of cars and main wiring in the house. The feeder cable is therefore an important part of the aerial installation, and for best results must be matched to the electrical characteristics of both the aerial and the receiving set. This is done

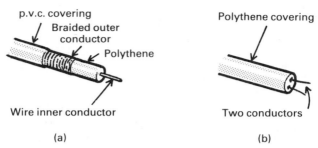

Fig. 127 Construction of a typical co-axial feeder is shown in **(a)**, a balanced line appearing as in **(b)**. Screened balanced cable is similar to **(b)** but has a braided outer screen and covering as in **(a)**

by comparing the electrical impedances of each item and matching the values as exactly as possible.

The impedance at the centre of a half-wave dipole aerial is about 72 Ω, and in this country feeder cables are designed to match this value. The two common types of cable are the twin, balanced cable and the co-axial cable. Fig. 127 shows these two types of cable. The impedance is a function of the diameters of the inner conductor and the inside of the outer conductors in the case of the co-axial feeder, and a function of the conductor diameters and their spacing in the case of the balanced feeder. The co-axial feeder is mostly used because it is naturally screened by the earthed outer conductor, is less affected by local interference and is more robust. The balanced feeder is more liable to damage, its characteristics are liable to vary with a wet surface in bad weather and it is affected by local interference unless it has a metal outer screen like the co-axial, but this makes it much more expensive.

The next important characteristic of feeder cables is attenuation, which is loss of signal strength due to transmission losses in the cable and which increases with frequency. It is expressed in decibels (dB), which is a unit based on the smallest sound audible to the human ear. It is used as a ratio of signal strength, based on voltage or power, between noise or interference and required signal, or, in the case of a cable, between signal strength at the receiving set and aerial connections, and it is also applied to loss and gain of signal strength in aerials, amplifiers and other audio/radio equipment. Tables are available that show numerical values of voltage and power loss or gain against decibel values.

Noise, as distinct from other forms of extraneous interference, is inherent in all audio equipment – in particular, the receiving set – and is apparent as a hissing background to the required signal from the loudspeaker. The signal/noise ratio must be kept as high as possible and, for ideal conditions, should approach 50 dB.

Aerial accessories

Needless to say, outdoor aerial fixings and aerials must be rigid, secure and strong enough to withstand high winds, snow and ice. The position of the aerial is important for maximum signal strength; therefore, if this is affected by weak fixings or weather conditions, the value of the aerial will be lost, and a loose aerial will cause variations in picture quality as well as signal strength.

Equipment for use with aerial feeders are available, including transistorised masthead amplifiers, power units and attenuators for controlling or regulating the aerial and feeder performance, in-terference suppression devices, matching or coupling transformers, diplexers and triplexers for connecting aerials of different band-widths to a common demand to ensure proper matching in each case, and star resistor networks and amplifiers to enable several sets to share the same signal from a common aerial. Nearly all these aim to match impedances and to maintain signal strength under different conditions of use.

The television set is often needed in different rooms. If separate aerial feeders are to be connected in parallel, matching pads or splitters must be used to maintain the same impedance matching for each outlet. But this is not necessary if co-axial cable extension leads are used from the socket in room 1 to a socket in room 2, and possibly again from room 2 to room 3, each cable being plugged into the feed socket of the previous lead. This avoids mismatching.

Provision for the aerial download is not always made in housing, but the unsightliness of aerial leads trailing down roofs and stapled untidily to outside walls and window frames should make such provision a necessity. A 20 mm plastic tube should be run as direct as possible from the aerial position, generally on the highest chim-ney stack or in the roof space, to the receiver position. If exposed above the roof tiles, the conduit should terminate in an inverted 'U' and bushed end to prevent the ingress of rain. At the receiver end, the conduit should terminate in a standard outlet box for a radio socket or sockets, as shown in Fig. 128; these are designed for the connection of co-axial cable and should be located as close as possible to the position of the receiving set. If desired, the conduit can pass through one or more similar outlet boxes, possibly located in first-floor bedrooms for alternative use on its way downwards. If

Fig. 128   Double T.V. co-axial socket outlet for direct connection to two separate T.V./f.m. co-axial downleads

(*M.K. Electric Ltd.*)

these provisions are made when a house is built or before a later redecoration, the conduit can be buried or otherwise concealed without unsightly wiring.

Aerial reception is seldom required for ordinary domestic sound radio reception in good signal areas because most receivers are extremely sensitive, portable, battery-operated transistor sets which do not need external aerials, except in remote areas. But Hi-Fi (popular name for high fidelity) radio and television receivers for the v.h.f. and u.h.f. wave-bands usually require an aerial and a mains supply.

It is best to position television sets for viewing where the daylight from windows or artificial lighting is not reflected by the screen, thus eliminating the need for excessive picture brightness for adequate visibility. A mains outlet socket should also be conveniently located near the aerial socket for connection of the set mains lead.

Earthing

The long-wire or long vertical rod aerial with a single conductor download requires a good receiver earth connection for efficient oscillation to occur between the two poles (aerial and earth) of the resonant circuit through the coupling coil of the receiver. Indirect earthing through the receiver chassis or mains supply is very inefficient, although often quite effective in areas of good signal strength. The earth lead should be a stout copper wire of the shortest possible length.

Modern receivers are not always isolated from the supply by a double-wound mains transformer, and the chassis may be directly connected to the neutral line of the supply; further, the aerial system is only isolated from the set by series capacitors. Consequently, a dangerous 'live' aerial at mains voltage may occur in the event of a fault or breakdown of certain components in the set. It is most important, therefore, to earth properly all exposed metal parts and the outer conductor of the aerial feeder cable, and to make sure that the supply neutral conductor is connected to the chassis if there is no mains transformer. It is also important to disconnect the set from the supply when adjusting the aerial or inspecting the chassis. In a.c./d.c. sets the chassis may be live and may have a separate earthing terminal.

There is something to be said for disconnecting the aerial lead from the receiver during a thunderstorm to avoid damage to the set in the event of lightning striking the aerial, but not for earthing the aerial instead, as this would turn the aerial into a lightning conductor and might result in much more serious damage if struck.

Receivers

The use of transistors and printed circuits has revolutionised both radio and television receiver design but, although transistors have increased the robustness of circuit construction and reduced the heat to be dissipated compared with that liberated by the thermionic valves in older sets, there is still a risk of overheating and possible fire if sets are not carefully positioned where there is free ventilation around the cabinet and no likelihood of curtains accidentally covering the set.

Stereophonic broadcasts are transmitted on Band II wavelengths, and the normal Band II aerial is used, but there is some loss of power and additional aerial gain may be required in weak signal areas. A 'mono' receiver will pick up the transmission, but if stereo reproduction is required a decoder will be necessary in the receiving set to separate the A-B sidebands of the carrier wave.

Colour television is transmitted on Bands IV and V on u.h.f. waves in most areas and, of course, requires a special colour receiver. Colour television sets are expensive – several times the cost of monochrome sets – and repairs and replacements are also very much more expensive, so colour television hire on a rental basis may be more attractive than purchase, especially where the charges are not much greater than those for hiring a monochrome set. Bands IV and V aerials may be used, but they need to be more responsive at the edges of the bandwidths to cope with the extra sub-carrier frequencies involved and to ensure good colour rendering. In general, colour television requires a better standard of aerial design and installation than for monochrome reception to ensure higher signal level and adequate bandwidth response, especially where signals are not strong. Colour television sets are also sensitive to masses of metal, such as radiators, in the vicinity of the set and this may affect the choice of location in a room.

A common feature of modern television receivers is remote control without any wires to the set, which is based on ultra-sonic or infra-red technologies. The hand-held control unit has buttons which take the place of all the controls normally on the receiver cabinet and control is effected up to an average room length away from the set.

Monochrome television sets have a loading of about 150 W and colour sets have a loading of about 350 W but generally have a high starting current when the set is first switched on which might blow a 3 A fuse in a 13 A plug. If this occurs the plug top fuse should be replaced by a 13 A fuse. This is an exception to the general rules on fusing which should otherwise be strictly adhered to. A hi-fi stereo receiver with record-player and loud speakers takes 100–120 W; but tape recorders and separate record-players only take about 40 W.

Video recorders

The latest development in domestic entertainment has been that of video recorders, the larger version of the audio recorders that are commonplace and can be carried in the pocket.

The video recorders now available are either full-sized units intended for use with a television set or miniaturised portable units intended for use with a small portable video camera.

The television version comprises all the tuning and station selection circuits that are in the normal television set, a tape recorder of high quality with 'read' and 'erase' heads, but without any display tube or audio sound circuits. The recorder has to be pre-tuned (as is the television set) for the desired stations and can then be switched on manually or by time switch to record any programme onto a cassette of magnetic recording tape inserted into the recorder. It is important that the video recorder is exactly matched and compatible with the television set with which it is to be used and in particular that it operates on the same broadcast signal characteristics as the set. Pre-recorded tapes can also be played back on the recorder and libraries of these tapes exist for sale or hire and public libraries lend them in nearly all towns.

This equipment enables one programme to be viewed on the screen while another is being recorded, and programmes can be recorded automatically when broadcast and viewed at a more convenient time. As with an audio recorder, when a programme is no longer required, it can be erased and recorded-over on the same tape many times before it is mechanically or electronically worn out.

The small portable recorder uses a miniature tape, is only tuned for the signal given by its associated video camera and records this continuously when the camera is running. The tape can be rewound and re-run in the camera to see the recording or can be erased and recorded-over in an identical manner to the larger units. The small video tapes are usually provided with a transfer or adapter cassette so that they can be played directly in a large television type recorder and viewed on a normal television screen or can be played back into the large recorder onto a normal sized cassette.

The cost of a high quality television recorder can be higher than the television set itself due to the precision tape recording mechanism and the cost of the portable recording units are many times the cost of the static units due to the miniaturisation and the electronic/optical systems needed for the camera.

Satellite television

It is now possible for private television broadcasting companies to use rockets launched into space to carry communications satellites capable of receiving, amplifying and re-transmitting radio and television signals. It is also possible to position the satellite accurately in geostationary orbit where it rotates with the earth and appears to be fixed in space when 'viewed' from a ground station.

A satellite aerial can cover a very large ground area from its elevated position and as it appears stationary can be designed to reflect television, communication and data signals at high power by means of a focussed narrow radio beam from a parabolic dish aerial at a ground station. Amplifiers fed from solar cells on board the satellite can further boost the signal received from the ground and small dish aerials on the satellite can rebeam the signals to ground receiving stations many miles from the original transmitter.

In time satellites belonging to companies in many countries will be beaming countless programmes to receivers at ground level. Unfortunately, there is no international agreement set up in advance to standardise signal characteristics so viewers will need to set up as many aerials and converters as there are different signals in order to receive them all. Many of the signals will originate from commercial stations who may charge the viewer for the programme so necessitating additional metering or pre-payment equipment to be accommodated in the home.

One solution to these problems might be a return to cable television where one central receiving station in a town or area has high quality aerials and amplifiers for all signals and relays these by cable to the viewers. Each house would then only require one small co-axial cable entering the house to enable all programmes to be received. This system would at least avoid a forest of dish aerials that would otherwise be required.

Interference
Radio interference is a very unpleasant nuisance and is heard as a continuous low-frequency noise on the radio set: the transmission being received at the time is usually completely drowned if the source of the interference is nearby. The most common source used to be the unsuppressed disturbance from car sparking plugs but nowadays cars are generally fitted with interference suppression devices. Today the most usual source of interference is faulty domestic appliances in which the interference suppression device, which must be provided where necessary by law, has broken down. Another possibility is the local electrical enthusiast or radio amateur who may be experimenting with apparatus which is not interference suppressed or which emits interfering radiation. All equipment which is liable to cause such a nuisance should have suppressors fitted; this includes fans and other motor driven appliances with series wound motors which have commutators and sparking brushes, and any other equipment which produces sparking or has inductive circuits that cause radiation. A room lighting switch that sparks on breaking the circuit can be heard on a radio set but this is not considered as radio interference because it is not continuous and it only affects a set in close proximity to the switch.

If interference is heard on a set it is often possible to trace the direction from which the interference comes because the aerial incorporated in the set has directional qualities which help to track down the location of the offending nuisance if the radio set is a handy portable one. If interference is serious and cannot be traced easily, British Telecom can investigate and locate the trouble; a leaflet on this subject is available at all post offices.

The theory and construction of radio receivers, amplifiers, tuners, loudspeakers and other electronic equipment for hi-fi sound reproduction are specialised subjects beyond the scope of this book. Suffice it to say that the electronic circuits in the receivers amplify and convert the radio signals to audible and visual effects in the loudspeaker and television tube respectively.

Telephones

The modern home should have provision for a telephone service which avoids the unsightliness of telephone wires stapled to window and door frames, skirtings and walls. The simplest answer is a concealed conduit from a point in the front of the house, outside where the telephone cable enters, to a position inside the house where the telephone will be located. British Telecom engineers can then pass their wiring through the conduit, and the good appearance of the exterior and interior of the house is maintained.

The telephone can be in any room, and the conduit, which may be steel or plastic, should terminate in a standard flush outlet box provided with a suitable overlapping cover plate for the telephone connection. The old method was to have a connector block inside the box and a bushed hole in the plate for the flexible cord as shown in Fig. 129(a), or a fixing for a surface type component for surface

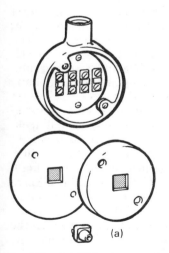

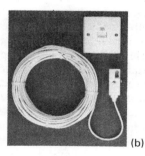

APPROVED for use
with telecommunication systems
run by British Telecommunications
in accordance with the conditions
in the instructions for use.

(c)

Fig. 129 (a) Telephone outlet box with flush and surface lids and rubber bush for cord (old method)
   (b) New Telecom socket on plate or box lid
                                            (*M.K. Electric Ltd.*)
   (c) British Telecom Approval Label for equipment

wiring. With all new installations, a British Telecom socket will be fitted which will enable subscribers to plug in their own telephone instruments independently. This unit is shown in Fig. 129(b) and is designed to encourage the installation of several different telephone facilities such as message recorders or facsimile machines. The equipment can then be bought from British Telecom or other approved makers in numerous styles and colours to suit subscribers' preferences.

It is now common to have an extension telephone in another room – for example, a study or bedroom – in which case the empty conduit should extend from the outlet box at the first position to a similar outlet box in the second position. More than one extension can be provided if required, for which the conduit is extended in the same way. In such an arrangement the telephone points are usually connected in parallel and, if required, each instrument can have a push button to mute its own bell. There are also more elaborate systems with central switchboards, but these are not usually required for domestic use.

Extension telephones or telephone socket outlets are a great convenience and can be provided by British Telecom or installed by the user using simple plug-in wiring kits. The kits and additional telephones may be purchased from British Telecom or a number of specialist manufacturers but must all be marked as British Telecom Approved and no alteration to permanent wiring is permitted. The telephone system voltage is low and wiring with low-grade insulation is used. With the rapidly developing systems of interconnection it will eventually be possible to dial a call on any telephone instrument for connection to any part of the world.

The development of computers, micro-circuits and silicon chip electronics is revolutionising communications theory and techniques. Telephone exchanges now house banks of micro-circuit boards without the many moving parts of the old exchanges which can become faulty and need much maintenance. Computer technology enables them to do all the necessary selection and connection of lines. Optical cables with numerous glass fibre cores, very much smaller than the present metal conductor cables and capable of carrying very many more signals with light radiation through the cores, will give the system almost unbounded possibilities in the future.

Already a Viewdata service is available to all telephone subscribers called Prestel, which only involves the addition of a special selector and code unit linked to the telephone line and the television screen to enable the user to call up on to the television screen any information required from a central data and information computer bank. This involves using the existing television set, and not a special new one. Television sets can also be purchased with circuits capable of receiving a broadcast Teletext data system and computer owners can receive various data systems through their British Telecom telephone. The most common information required would probably be weather conditions in certain areas, train and air travel times and availability etc., but much more specialised and diverse information on all kinds of subjects that would otherwise have to be obtained from libraries, industry and expert authorities is available. Ultimately there will be almost no limit to information availability.

Future developments include visual appearance on a screen of the persons making and receiving a telephone call so that the speakers can see each other while conversing. This will also embrace conference facilities which will enable a meeting of several people to take place, each sitting in his own office or home, as though they were all together in one room. The numerous possibilities stretch the imagination, but some obvious uses will certainly be the monitoring of small children in nurseries, the remote control of automatic home equipment and security arrangements and centralised meter reading or recording of public utility services.

Although not usually required for home installations the entrance telephone is common in blocks of flats and equipment is available for viewing the caller on a miniature screen also. Verbal communication with a caller is simple and safe without going to open the front door. If it is desired to allow a caller to enter, an electric door lock is fitted with a control push or switch near the inside telephone which, when operated, unlatches the entrance door and allows the door to open.

Telephone instruments

The instruments used in telephone systems comprise the transmitter, which converts sound waves into electric currents for transmission over long distances; the receiver, which may be an ear-piece or loudspeaker for converting the electric currents back into sound waves; bells, buzzers or light signals for drawing attention to the need to answer a call; amplifiers and repeaters for loudspeaking telephones and long distance communication; and manual or automatic switching arrangements to connect callers to the required receiver.

THE TRANSMITTER Sound transmitters or microphones incorporate a diaphragm which is designed to vibrate in response to sound waves carried by the air and impinging on the diaphragm, so transmitting the frequency and amplitude, or pitch and loudness, together with the quality of the speech or music being communicated, to a very sensitive electromagnet or a mass of carbon granules in the space behind the diaphragm. The diaphragm forms the front of a rigid insulating box containing the granules, and the diaphragm and a backplate behind the granules form the two electrical connections, so that a current passing through the carbon granules is varied by the changes in resistance caused by the varying mechanical pressure on them as the diaphragm vibrates with the sound. A section of a typical microphone is shown in Fig. 130. There are various designs of microphone, but they are usually constructed as a capsule, complete with terminals, which is encased in a telephone handset.

A telephone induction coil is a transformer with an open magnetic core of soft iron wires, with the primary winding in series with a battery and transmitter, and the secondary winding to step up the voltage for the transmission line to the receiving end, where long lines are involved.

THE RECEIVER   The telephone receiver consists of conical diaphragm attached to the end of an armature which acts as a lever, pivoted at its centre across the poles of an electromagnet, and vibrates the diaphragm through a pin fixed at one end of the lever, as shown in Fig. 131. This unit is encapsulated in a similar way to the transmitter for enclosing in a moulded handset. The electric current, corresponding to that in the transmitter, passes through the coils and causes the diaphragm to vibrate in unison, thus converting

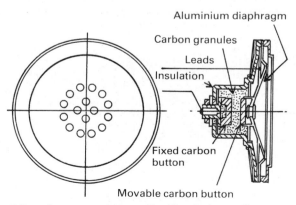

Fig. 130   Microphone capsule in section (leads shown diagrammatically)

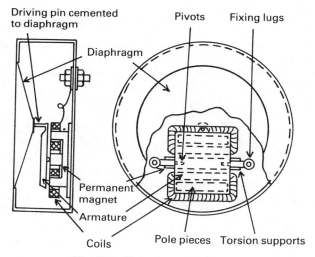

Fig. 131 Telephone receiver

the electrical energy back into mechanical energy and the varying current into sound waves, which reach the ear.

The simplest form of communication system which is often used in the home is for nursery calls whereby a crying baby and a microphone in the nursery can be heard on a loud speaker in the living room; but it is so convenient to have a conversation between ground floor and a distant floor or room without having to traverse the distance between in a large house that a small private telephone system will now be considered.

The telephone circuit

A simple telephone circuit for the house with two stations is shown in Fig. 132 (page 242). Automatic switching is obtained by hanging the handset on a lever which operates the two-way switch, leaving the bell circuit in series with the distant station when out of use and putting the handset in series with the distant station when it is lifted off the hook; when the call push is operated before lifting the handset it rings the distant bell. This is suitable for domestic use but for long distances induction coils would be connected in the circuits, as in Fig. 133 (page 242). In a domestic circuit, very little power is required; a few Leclanché or dry cells are adequate to overcome line and instrument resistances. In large installations, however,

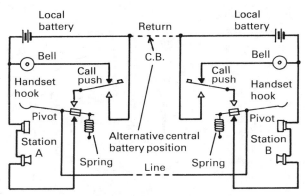

Fig. 132   Simple telephone circuit

batteries of secondary cells up to 100 V are often used. A d.c. supply
is, of course, necessary for a telephone circuit.

Wiring sizes are related to voltage drop and circuit resistances,
but in small domestic installations 1·0 mm² or 1·5 mm² copper bell
wiring can be used. Where part of a telecommunication system is
directly connected to the public electricity supply, that part must
comply with I.E.E. Regulations, i.e. the supply to a double-wound
transformer providing extra-low voltage to the system (with conversion to d.c. as for battery charging), otherwise the Regulations do
not apply to other parts of the system not directly connected to the
mains. In the best installations concealed 16 mm or 20 mm conduit
would be used for wiring, but as twin bell wire is so small and not
very noticeable if fixed with insulated staples on surfaces in a neat
and inconspicuous manner, it is generally installed in this way if a

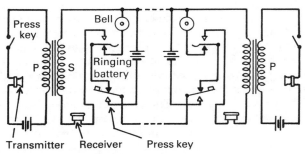

Fig. 133   Simple telephone circuit with induction coils

higher standard is not called for. This wiring must be completely separated from any other mains wiring.

The telephone circuitry and components described are common to all old and many existing telephone systems but electronic developments have made the control and switching arrangements by mechanical means out-of-date. Transistors and silicon chips are now made to carry out these functions and occupy much less space, so that new telephonic equipment will be quite different, although the circuits in principle will be the same and handset contacts will probably be retained.

Loudspeaking telephones
With the development of transistor circuitry, there is a trend towards the use of loudspeakers for telephone receivers and sensitive microphones without handsets, but there will always be a need for the handset telephone instrument where private conversation is necessary. This trend, however, has many advantages: the microphone and remote loudspeaker are ideal for monitoring baby welfare in the nursery; they also allow freer movement and keep the hands free for writing while conducting a conversation.

Loudspeakers are simply a larger version of the electromagnetic receiver, having a large conical diaphragm to move a greater volume of air; this requires more power to operate than the handset receiver and therefore amplification of the signal is necessary. Before the advent of the transistor, this meant expensive equipment with thermionic valves which made it generally impracticable, but the comparative cheapness and small space requirements of transistors and printed circuits have made amplification of very small electric currents feasible, especially for audio purposes.

Bells and alarms

The generic term for bells, buzzers, sirens, hooters and other forms of alarm, in the sense of calling attention to something or need for action, is 'sounder', which is now adopted accordingly in official and technical references to them, but for the purpose of this book we will use the well known common words and start with 'Bells'.

The d.c. trembler bell is a simple piece of electromagnetic apparatus; the working principle is shown by diagram in Fig. 134 (page 244). The electromagnet consists of a soft iron core, C, which is easily magnetised when the current traverses the magnet coils, MM; but when the current ceases the iron loses its magnetism; the tip of the screw makes contact with the spring, S, and the circuit is completed

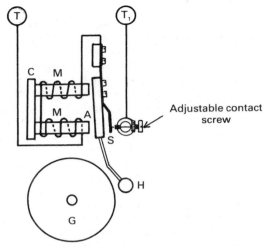

Fig. 134   Internal connections of electric bell

by pressing a bell-push. The current through the winding, MM, causes the iron core to be magnetised and the armature, A, is attracted. This breaks the contact with the spring, S, the current in the coil ceases, the magnets lose their magnetism and the armature flies back to repeat the sequence. This gives a rapid trembling movement to the hammer, H, against the gong, G. Various types and sizes of gong, either of bell metal or steel wire, can be obtained to give distinctive notes; but if a quieter note is required, a buzzer can be used. Sources of trouble in a bell or buzzer are generally due to a fault in the contact-breaker, caused either by weakening of the spring or by corrosion of the contact screw.

A buzzer works on the same principle as the trembling bell, but the hammer and gong are absent. The vibration of the armature, which is of different design and relatively light, gives the characteristic high-pitched note.

Bells for a.c. are designed to vibrate with the supply frequency (50 Hz) and do not need a contact-breaker but an ordinary d.c. bell will usually work well on a.c. also.

Bells and buzzers of the plastic-moulded pattern are common and are usually made suitable for extra-low voltage operation. Mains voltage bells are available but not usually provided for domestic premises. Small double-wound transformers supplied from the mains are commonly used for domestic bell circuits.

Electric bells on a.c. supplies

The magneto bell is operated on low-frequency alternating current and its use is confined to telephone installations. The construction is similar to that shown in Fig. 134, but the contact-breaker is omitted; the two ends of the magnet coil windings are taken directly to the terminals, T and T; the armature, A, forms a rocker pivoted at its centre; and the hammer, H, vibrates between two bells with the supply frequency. With 240 V, 50 Hz supply a bell transformer is generally used for the bell circuit with extra-low voltage output for the bell. They are normally provided with terminals for secondary voltages of 4, 6 and 8 volts for the bell circuit. *This should be of the double-wound type*, in which the two windings are entirely separate and insulated from one another, and, if not double-insulated, have an earthed shield between them or be mounted on separate limbs of the core which must be earthed. Fuses should be fitted to protect the primary (240 V) winding, and one terminal of the secondary winding and the iron core should be earthed if a metallic shield is not fitted between the windings. With two bell-pushes, one at the front door and the other at the back door, the low-voltage winding can be connected via the respective pushes to a bell and a buzzer instead of using an indicator, as illustrated in Fig. 94, p. 146, where the bell transformer, bell and buzzer are shown mounted above the kitchen door.

It is important to ensure that the bell pushes are connected in the low voltage circuit winding and not in the 240 V winding of the bell-transformer. The operation of the bell would be identical but a very dangerous, potentially lethal, situation would exist at every external bell push. Several serious accidents have occurred as a result of this simple error.

Continuous-action bells

These are bells which continue to ring after the distant bell-push or contact is released and only cease when the operating cord is pulled or the local battery is exhausted. The local battery circuit is shown in Fig. 135 (page 246), together with additional mechanism. With the first movement of the armature, A, the trigger, U, disengages from the projection on the end of the armature and makes contact with the auxiliary pillar, X. The bell is then directly connected to the local battery. When the operating cord is pulled down, the trigger again engages with the projection on the end of the armature, the original connection is restored and the local circuit is broken. This type of bell is normally used for alarm purposes.

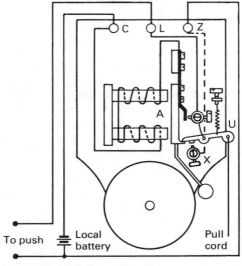

Fig. 135   Continuous-ringing bell

Chimes

A popular modern sounder for the house is the chime, which is a very simple device consisting of an electromagnet in series with the door-push and battery or transformer. When the circuit is closed by operating the push momentarily, the horizontal sliding magnet core of soft iron with nylon ends is drawn rapidly into the magnet coil against the pressure of a weak coil spring; the leading end hits the sounding metal and rebounds with the aid of the spring to hit another piece of sounding metal at the other end of its travel. The two pieces of metal are usually of different tones to sound like a chime and may be hanging tubes of different lengths or small strips of metal as used in a toy xylophone.

Bell-pushes

Bell-pushes of many types are available. For outside situations the weatherproof, metal-barrel type is best, while for bedside positions the ceiling pendant pear push or cord-operated pull type may be more convenient than the wall type and avoids the wiring having to be buried in the walls. For special purposes, such as burglar alarms, door, window and floor contacts are used, preferably with concealed wiring, but such items are usually provided and installed by specialist contractors. Fire alarm contacts which make the bell circuit when the front glass is broken are also a form of bell contact with reverse action, but fire alarms are not usually installed in the average house. Illuminated bell pushes are available which are very suitable for use at dark entrances, but the lamp in them is connected across the contacts and so becomes a continuous drain on the battery or other form of supply to the circuit. It is therefore somewhat wasteful as it is always in circuit 24 hours a day.

Relays

A relay is used to close a local battery circuit near the bell when the current from a distant push would not be strong enough to ring the bell, owing to the loss of voltage over a long run of small wire. Relays are also used with luminous call systems, the signal being sent by the distant push, which rings the bell, and the locking relay connects the local battery to the lamp, which remains alight until the relay is reset. The circuit diagram is shown in Fig. 136.

The relay consists of a form of electromagnet with an armature which is similar to a bell mechanism but designed to have a single movement and carries contacts that make or break circuit with fixed contacts. They are also used with mechanical indicators in large establishments with long runs, as in Fig. 137 (page 248), which shows four-way indicator circuitry with separate relays and a local battery.

Indicators

Indicators are provided to show from which room or place a bell has been rung. There are three types of movement: pendulum, mechanical and electrical replacement. There is also the luminous

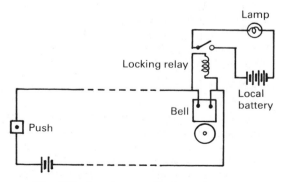

Fig. 136 Luminous call system with relay

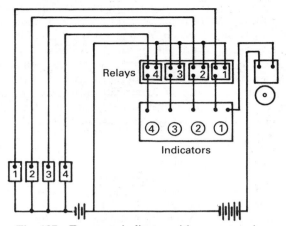

Fig. 137   Four-way indicator with separate relays

indicator. Of these, the first is the simplest and consists of a pendulum with an iron armature which is attracted by an electromagnet and starts swinging. This is illustrated in Fig. 138, which also shows the circuits using a bell transformer. The mechanical-replacement movement allows a flag to drop and become visible and has a resetting rod projecting from the side of the indicator case, while the electrical-replacement type is similar but the flags are replaced by an electromagnet and a push-button, which can be located either adjacent to the indicator or at a convenient point some distance away. Some indicators include a bell, which is mounted on the base of the indicator. Luminous indicators employ single or groups of lamp signals connected in a similar way to that shown in Fig. 136, page 247. They are particularly useful where silence is required – for example, in hospitals – but can be used in conjunction with audible signals if required. When purchasing any of the bell equipment described it should be clearly stated whether it is for a d.c. or an a.c. supply.

An electronic method of indicating numbers and letters is effected by using a liquid crystal display. An arrangement of liquid crystal cells, so placed to form any number or letter required by selection when the circuit voltage energises each segment of the shape required, reflects ambient light or transmits a light source behind them as a result of changes in the molecules of the crystals. This is used in wrist watches and pocket computers and many other applications in which a very low luminous intensity is satisfactory.

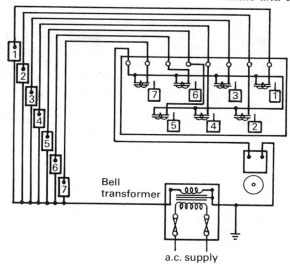

Fig. 138 Bell transformer and indicator circuits (supply neutral not earthed)

Bell wiring

Good materials and careful installation are essential if the bell system is to be reliable. Concealed conduit or surface wiring can be used. Similar general considerations as for light and power wiring apply. Wiring Regulations, however, do not apply to installations in which the wiring is not directly connected to the public supply mains, and for the extra-low voltages mentioned lighter, smaller and cheaper materials can be used. No wiring should be buried solid in walls without protective tubing. The size of copper wire should be not less than 1/0·85 mm; single p.v.c., lightly insulated, in seven different colours, and twin bell wire can be used for internal surface wiring. Kinks in the wire must be avoided, otherwise the single-strand conductor may break later. For the best-quality work, especially in damp situations and outdoors, twin p.v.c.-sheathed wire should be used. For such situations and larger section wiring where voltage drop has to be allowed for in long runs, mains voltage cables as used for lighting are often employed. Plastic clips which do not cut into the insulation are preferable to insulated staples; but if staples are used they should have an insulating saddle made of fibre to protect the wires. It is important to draw a circuit diagram to insure that 'lead' and 'return' wires will be provided to suit every circuit operation envisaged.

Bell wiring must be kept separate from lighting and power circuits, and must not be run in the conduit carrying these circuits.

There are many other arrangements of bell circuits designed to suit special requirements, and when there is a considerable distance between bell and push an 'earth return' is sometimes used to economise on wire, using earth plates or water pipes for the return connections. Gas pipes must never be used for earthing connections.

For signalling purposes, single-stroke bells and buzzers are used; special contacts which operate like a Morse key can also be used, and such circuits often employ a single-line wire and earth return.

Fire Alarms

In recent years, fires in houses have resulted in serious injury or death of the occupants not only due to burns but more often the result of asphyxiation from smoke, particularly from burning upholstery which contains polystyrene filling and which gives off a poisonous and voluminous smoke under combustion. Consequently the early detection of the outbreak of fire in the house has become of vital importance to householders, who have been strongly advised by Fire Brigades to install smoke detectors to give warning of an impending outbreak of fire, and to give time to escape, to summon help or to quench the source of a fire before it becomes serious.

In addition, there has been a large campaign in the U.K. to fit new windows to domestic premises, some with double glazing, and many windows have become large sheets of glass or 'Picture' windows without any means of escape in the event of fire in the house. In addition, burglary has increased and windows have been sealed against intruders. As a result, smoke detectors, which have always been available for business or industrial premises and public buildings, are now marketed for domestic use, each with a loud alarm sounder and battery operated by means of circuitry and a small radioactive source which can give an electrical charge to any tiny particle of a combustion product such as carbon, and this charge can then be detected. These units are known as 'Ionisation' detectors and are more sensitive than optical detectors that rely upon smoke obscuration of a photo-electric cell.

The units are completely self-contained; their use is recommended in every home, and the only care that needs to be exercised is in disposing of the unit at the end of its working life. The units must not be incinerated or crushed as this could release a small amount of a radioactive substance; they should be disposed of through the local refuse service who will collect the units and safely dispose of them, or by returning them to the manufacturer. Full instructions for disposal are given on the units and should be followed to the letter.

Smoke detectors can be obtained easily, are not expensive and can be installed by the householder. For the common semi-detached or terraced six-roomed home, at least two units fixed to ceilings or high on walls are advisable (one in the ground floor entrance hall and one over the first floor stairs landing, not too far to be inaudible in occupied rooms), though one unit in the entrance hall makes a useful start for a smoke detector system which can be added to later. Instructions for siting and fixing are usually supplied with these units when sold. A typical 6 ins square, 2 ins projection smoke detector is shown in Fig. 139.

Burglar Alarms

A variety of burglar alarms and intruder detectors are now available to the general public whereas previously these had only been available to specialist contractors for fitting to large houses with valuable contents, or to prevent trespassing.

The units generally use electronic circuitry and ultrasonic technology for detection. There is always the chance of a false alarm sounding the loud alarm bell, klaxon or other warning device which is often fixed outside the house and can be activated by harmless noises or sounds from vehicles or church bells if loud enough. A common type of detector transmits an ultrasonic beam over the protected area which, if interrupted by an intruder, initiates the alarm. (See Fig. 140).

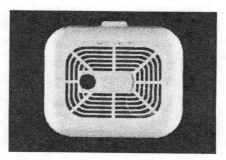

Fig. 139   Smoke detector fire alarm

(*Black & Decker*)

Another form of detector projects low intensity pulsed ultrasonic signals into a room which reflects a pattern of signals back into the receiver. If something changes position in the room, the detector will respond after a few pulses have established a new pattern and sound the alarm. This type of detector has the advantage that it can be set to have variable sensitivity to transient sound and can ignore things that constantly change such as curtains flapping or a cat walking across the room. If however an object is permanently removed or added, or if the cat sits down for a minute, this may set off the alarm.

Other forms of detector are the old-fashioned concealed window and door contacts, or a rubber mat with electrical contacts which raise an alarm when trodden on, and the visible or infra-red light beam across a passage way to a light sensitive cell which sets off an alarm when interrupted by a trespasser. These types of detector usually have a separate battery and permanent, concealed wiring between the units.

Static

Since the introduction of nylon and other man-made fibres into the weaving industries, their use for carpets has become widespread and carpets of standard rectangular dimensions have been largely out-moded by carpets fitted from wall to wall in rooms and corri-

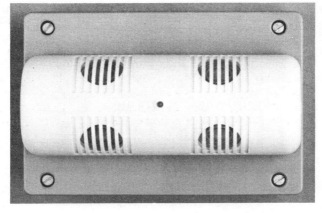

Fig. 140　Ultrasonic presence detector
(*M.K. Electric Ltd.*)

dors. As a result the occupants may be walking on wool or synthetic fibre carpet without change all the time they are indoors. Slippers, shoes or bare feet frequently meeting with friction on carpets have similar effects to the old experiment of rubbing amber on silk to produce sparks of static electricity. The occupant becomes charged with electricity in the same way and to a varying degree; and only becomes discharged by gradual leakage over a period or suddenly when an earthed conducting object is touched. In the former case nothing abnormal is felt but in the latter it is possible for a most unpleasant shock to be experienced measuring several thousand volts, though the current would be infinitesimal. This has been a fairly common experience in large, modern, open-plan offices in tall city buildings with dry (low humidity – less than 30%) conditions where the first contact with earthed metal was, perhaps, the stair handrail or an earthed metal switch or socket outlet plate. Such electrical shocks experienced by office staff led to suspicions that the electrical installation was at fault, causing exposed metal to become 'live'. Such effects are not generally found in the home but in large houses having fully carpeted rooms, and low humidity with central heating it is possible that sensitive people may be affected in the same way. Static has also been observed with nylon underclothing which sparks when stripped from the body and it has also occurred when separating electric blankets from sheets on the bed.

The remedy is to prevent these effects by causing the static electricity that builds up on the carpet surface and on the body to leak away readily. The ideal atmospheric condition for this to happen naturally is about 65° relative humidity but otherwise the simplest method of dealing with the problem is by treating the carpets with an anti-static or humidifying agent which is sprayed on to the carpet and left to dry off overnight. Air conditioning can, of course, provide a remedy by controlling humidification to ensure an adequate level of moisture in the carpet.

It should be added that rubber and p.v.c. floor coverings are not immune from static, and different floor polishes can effect a cure: but trouble of this kind has only so far been found in hospital wards.

13

Electrical Safety and Treatment for Shock and Burns

Although the safe use of electricity has developed rapidly since the beginning of the century, electrical accidents are still too frequent and serious to neglect some mention of further safety considerations, even repeating those specific safety requirements already covered in previous chapters, as well as how to deal with electric shock.

Safety

First, the provisions for safety made by the electrical industry should be recognised. The Institution of Electrical Engineers publish their *Regulations for Electrical Installations*, which are designed primarily to ensure safety from fire and shock, and are accepted throughout the industry as the main guide to good practice. These ensure good insulation, construction, earthing and installation methods, which only need the addition of good workmanship to make an installation quite safe to use. Good workmanship is encouraged by having a National Inspection Council with a Roll of Approved Contractors, who undertake to carry out safe installation work and the employment of graded and qualified electricians.

In addition, the user must be discriminating in purchasing good quality electrical appliances. To help him there is a national organisation, the British Electrical Approvals Board, which is supported by the Electrical Industry and the British Standards Institution and is intended to test and approve domestic electrical appliances, both imported and made in this country, for safety. Therefore any appliances or equipment bearing the B.E.A.B. Kitemark with B.S. 3456 for domestic appliances or 2769 for tools, shown in Fig. 141

Fig. 141 B.E.A.B. mark

can be relied on for safe use. The materials and equipment used in wiring installations are covered by various British Standards which ensure the safety, suitability and adequacy of these items, and Standards or Codes of Practice cover proper installation design for Artificial Lighting; Floor Warming; Domestic Electric Water Heating; Safety of Household Electrical Appliances; Electric Luminaires; Lightning Protection; Earthing; Reception of Sound and Television Broadcasting; Radio Interference Suppression and Telephone, Telegraph and Data Installations, and Safety of Domestic Sound and Vision Equipment, among many other issues.

In the past, many imported electrical appliances and toys were poorly constructed and dangerous to use, and some have caused fatal accidents. One of the most dangerous hazards has been the colouring of flexible lead cores when Continental standard colouring was different from British standard colouring, but this has been overcome by the adoption of International standard colours, which is now compulsory under the Consumer Protection Act, 1961. It is therefore most important for safety to know the correct colouring, as shown in Figs 53 and 54 (page 82). In this respect the following safety hints should be followed:

1. Follow the directions or polarity markings in the plug.
2. If the flexible lead core has other than standard colours, do not connect it without advice from a qualified electrician, or an electrical shop or showroom.
3. Never use a two-pin plug for a three-wire flex.
4. Never use the earth terminal when connecting a two-wire flex to a three-pin plug.
5. If the appliance has a metal case, a three-core flex and three-pin

plug must be used, but this is not necessary if the appliance is double-insulated and marked ▣.

In using electrical appliances, it is important to remember that heat from the appliance must be allowed to escape, and that restriction of ventilation or cooling will cause the appliance to get overheated and may result in a fire. This applies to all electrical equipment, especially T.V. sets and luminaires. Even electric blankets will cause fires if covered by too much for too long. All equipment with exposed live electrical elements, such as electric fires, must be properly guarded.

Flexible leads must be replaced when badly damaged or when the insulation is hardened by age. Long trailing flexible cords are a source of danger, too, and should be avoided if possible. Extension leads should be fitted and connected properly so that only protected sockets (not projecting pins) of couplings are live when plugged into an outlet. Flexible leads may be damaged if laid under lino or carpets, if passed through doorways or windows or compressed under the feet or legs of furniture and fittings. Flexible leads should not be stapled to a wall or skirting; damaged flexibles should be replaced as soon as possible, and never joined if broken except with purpose-made couplers in the correct way (as referred to above).

Always remove plugs when cleaning portable appliances, or switch off the control unit in the case of a cooker. Never take a portable appliance into a bathroom other than one that can be plugged into a proper shaver supply unit if installed.

Make sure that any portable appliance is either all-insulated (with no exposed metalwork that can become live in the event of a fault) or, if not, that it is properly earthed. Earthing of the metalwork in an appliance ensures that it cannot become live at a dangerous voltage (see page 106).

Ensure that no appliance is over-fused and that it has the correct cartridge fuse in the plug. When purchased, 13 A plugs are usually fitted with a 13 A fuse but this should be removed and replaced with a 3 A fuse if the appliance it is required to connect has less than 750 W marked on the nameplate.

All radiant fires should be adequately guarded to protect children, in addition to the guard in front of the elements on the fire itself. There are serious risks in controlling radiant fires with time switches because of possible handling by a person who believes the fire to be 'dead' and it is automatically switched on while being handled.

Always dry wet hands before touching or handling electrical

appliances, switches or plugs. Never use a higher wattage lamp in a light fitting than it is designed for because overheating may occur and lampshades made of flammable material too close to a hot lamp may burn. Never use a knife or fork to remove bread stuck inside a toaster before unplugging the toaster from the socket. Never fill a kettle from the tap without unplugging it from the socket first. Similarly, unplug a steam iron before filling with water and keep the outside of the iron dry.

A flex-holder on the ironing board contributes to safety; and never coil the flexible on a *hot* iron. Never use a lamp-holder to connect an iron or any other appliance other than a lamp. Make sure an electric under-blanket is switched off before getting into bed; avoid folding or creasing electric blankets and keep them dry.

Television sets should be unplugged at night – not just switched off. It is also a sensible precaution to unplug all equipment not in use when going out, and to switch off your electricity supply when going away on holiday but don't forget the freezer may be full of perishable food!

All these precautions demand an electrical 'sense' but we are now in an electrical age and it behoves us to fit ourselves for the enjoyment of this, one of nature's gifts, without endangering life or limb in the process. Nevertheless, man is imperfect and mistakes will be made and accidents will happen, so we must conclude with some thoughts on what to do when such unfortunate circumstances occur.

Electrical accidents

This subject can be divided into:

    (a) rescue of a victim of electric shock;
    (b) treatment of shock;
    (c) treatment of electric burns.

All these emergencies require quick, cool and clear thinking. Ignorant and thoughtless action can be dangerous for all concerned.

Rescue

As the victim may still be in contact with live conductors, the rescuer must not touch the victim or he too may receive an electric shock. So, the first thing to do is to switch off or unplug the supply, if this can be done without delay. If not, the rescuer must either insulate himself from 'earth' by standing on an insulating material (e.g. thick rubber, dry wood) or by using dry rope or timber (e.g. a broom handle) to break the victim's contact with live conductors. The victim should not be moved but be made warm and comfortable, using blankets or additional clothing, until a doctor arrives. But the essence of treatment for shock is urgency, so no time must be lost before applying appropriate treatment because delay rapidly reduces the possibility of recovery.

Shock treatment
Treatment of shock is most urgent because shock has two critical effects: it may stop the heart's action directly or it may stop respiration, both of which must be restored as soon as possible if the victim is to recover fully. A common situation is to be holding a live appliance in one hand while the other hand touches earthed metal (possibly a water tap) or while standing on a damp concrete or tile floor. The effect of the electric shock is to make the forearm muscles contract and tighten the grip so that the victim is unable to let go (hence the importance of switching off the electricity supply quickly). Another effect of accidentally touching a live conductor is for the shock current to cause rapid withdrawal of the arm, and, in passing through the body, the back muscles contract and throw the victim backwards. The passage of shock current through the body from an upper limb to some other limb generally affects the heart and causes ventricular fibrillation (or disturbance of the normal action of the main pumping chambers of the heart) which is usually fatal. Respiration is affected if the shock current passes through the respiratory control centre at the base of the skull to one of the limbs, or the shock current may cause direct contraction of the respiratory muscles and result in asphyxia.

If the heart is affected, blood circulation can be maintained by cardiac massage – the rhythmical squeezing of the heart between breastbone and spine. This is difficult to do properly and is unadvisable unless the rescuer is an experienced and qualified first-aider; but a single, fairly heavy, blow with the flat of the hand to the lower breastbone might be sufficient to start the heart working again. The rescuer should be very sure that such treatment is necessary before attempting to carry it out because of the danger of breaking ribs which generally happens when the edge instead of the flat of the hand is used.

Artificial respiration
In the case of arrested breathing artificial respiration will often save life by introducing the necessary oxygen into the system and relaxing the respiratory muscles. It is clear that urgency demands immediate action in an effort to restore normal bodily conditions. As

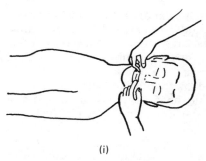

(i)

Remove false teeth, etc.

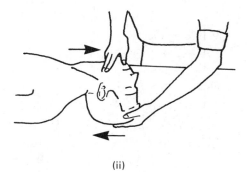

(ii)

Clear air passages, open mouth

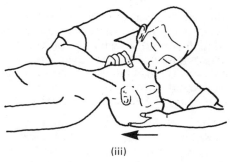

(iii)

Hold mouth open, and breathe in

Fig. 142 The mouth-to-mouth method of artificial respiration

soon as the victim is clear of contact with electricity, any false teeth, vomit, etc., should be removed from the mouth and artificial respiration started at once and continued until normal breathing resumes, even, if possible, during a doctor's examination and the dressing of wounds.

The approved method which has superseded older methods where there is no head injury is 'mouth-to-mouth' respiration, popularly called the 'kiss of life' which is illustrated in Fig. 142. The victim is just rolled flat on to his back. Then the air passages are cleared by tilting the patient's head back gently, the nostrils are closed by pinching with forefinger and thumb or pressing the cheek against them while operating; and, with the fingers of the other hand holding the chin of the patient from underneath to keep the mouth open, a deep breath is taken and the mouth placed over the patient's mouth so that there is no leak of air. Then, after breathing into the patient's mouth gently – especially so with children – until the chest rises, the mouth is taken away to allow air to be expelled and the process is repeated every five seconds. In the case of babies or small children the rescuer breathes into the nose and mouth (i.e. the nostrils are not pinched closed). The important thing about artificial respiration is to keep it up continuously until the patient has recovered normal respiration or until a doctor advises that treatment can cease. Colour returning to the face is an indication of recovery and is often evident – particularly with women and children – before self-breathing takes place. With the help of others in relay, a long period of treatment can be achieved, and, in the absence of expert advice, should be maintained continuously for at least 60 minutes or until normal respiration is restored.

Burns
Submersion of the affected part in cold water for a while will assist recovery, or covering the part with a sterile dressing or a clean handkerchief soaked in cold water until proper treatment by a qualified person is obtained. Medical advice must be sought if the skin has blistered or broken. Under no circumstances should oil, grease or powder be applied to the affected area.

Delayed shock
Physical shock and collapse can occur minutes after apparent recovery, so the recovering patient should be watched and cared for, with stimulants if necessary, for some time before being finally released but nothing should be given by mouth to a patient in this condition, neither medication, food or drink, until professional advice is obtained.

Index

Extension leads, 180, 256

Fan heaters, 155
Flexible cords
  cables, circuits and testing, 79–108
  choice of, 83
  connecting to plugs, 81, 82
  current rating of, 81
  for domestic appliances, 80
  three-core, 81
Floor heating and thermal storage, 160
Fluorescent lamp, 135
  circuits, 136, 137
Food processors, 179
Freezing cabinets, 196
Frequency
  radio and television bands, 223
  and standard voltages, 34
Fuses
  and circuit protection, 28
  cartridge for 13 A plugs, 30
Fuse wire for semi-enclosed fuses, 30

Garage equipment, 199–200
Garden appliances, 197, 200, 201

Handlamps, portable, 200
Harness wiring system, 73, 77, 217
Heat
  energy, 149
  requirements, calculation of. 151–154
  transmission factors, typical, 151
Heaters
  electric water, 168
  location of, 157
  tubular, 158
Heating, electric space and water, 149–176
Heat pump and solar heat, 165
Hot water
  conversion of existing system, 175
  household consumption, 175

House insulation, 167
Hydrometer, 50

IEE Wiring Regulations, 1, *see also* preface
Illuminance
  at a point, 131
  mean spherical, 130
Illumination
  calculations, 140
  intensities, 128, 138, 145
  units, 130
Immersion heaters, 172
Impedance
  aerial feeder cable, 231
  earth loop, 103
Incoming supply and methods of distribution, 36
Indicators (bell), 248, 249
Inductance (radio), 220, 223
Induction
  coil, 240, 241
  motor, 45
Installation
  layout of a typical, 205–219
  materials (typical house), 206
  systems, 66, 217
  testing, 95
Installation accessories
  for radio and television, 231
  and switching, 109–127
Instantaneous water heaters, 168
Insulation
  resistance, 98
  thermal, in the house, 167
Insulators, 26
Interference-radio, 227, 230, 236
Intermediate switching, 120
Intruder alarms, 251
Ironer, rotary, 192
Irons, electric, 187

Junction box, 73

Kettles, electric, 187
Kitchen
  layout, 146, 218
  lighting, 147

Lampholders, 116
Lamps
  effect of incorrect voltage, 36
  electric, 129